International Political Economy Series

General Editor: **Timothy M. Shaw**, Professor and Director, Institute of International Relations, The University of the West Indies, Trinidad & Tobago

Titles include:

José Carlos Marques, and Peter Utting *(editors)*
BUSINESS, POLITICS AND PUBLIC POLICY
Implications for Inclusive Development

S. Javed Maswood
THE SOUTH IN INTERNATIONAL ECONOMIC REGIMES
Whose Globalization?

John Minns
THE POLITICS OF DEVELOPMENTALISM
The Midas States of Mexico, South Korea and Taiwan

Philip Nel
THE POLITICS OF ECONOMIC INEQUALITY IN DEVELOPING COUNTRIES

Pia Riggirozzi
ADVANCING GOVERNANCE IN THE SOUTH
What are the Roles for International Financial Institutions in Developing
States?

Lars Rudebeck, Olle Törnquist and Virgilio Rojas *(editors)*
DEMOCRATIZATION IN THE THIRD WORLD
Concrete Cases in Comparative and Theoretical Perspective

Eunice N. Sahle
WORLD ORDERS, DEVELOPMENT AND TRANSFORMATION

Benu Schneider *(editor)*
THE ROAD TO INTERNATIONAL FINANCIAL STABILITY
Are Key Financial Standards the Answer?

Adam Sneyd
GOVERNING COTTON
Globalization and Poverty in Africa

Howard Stein *(editor)*
ASIAN INDUSTRIALIZATION AND AFRICA
Studies in Policy Alternatives to Structural Adjustment

William Vlcek
OFFSHORE FINANCE AND SMALL STATES
Sovereignty, Size and Money

International Political Economy Studies
Series Standing Order ISBN 978–0–333–71708–0 hardcover
Series Standing Order ISBN 978–0–333–71110–1 paperback

You can receive future titles in this series as they are published by placing a standing order.
Please contact your bookseller or, in case of difficulty, write to us at the address below with
your name and address, the title of the series and one of the ISBNs quoted above.

Customer Services Department, Macmillan Distribution Ltd, Houndmills, Basingstoke,
Hampshire RG21 6XS, England

Governing Cotton

Globalization and Poverty in Africa

Adam Sneyd

Assistant Professor, Department of Political Science, University of Guelph, Canada

First published 2011 by
PALGRAVE MACMILLAN

Palgrave Macmillan in the UK is an imprint of Macmillan Publishers Limited, registered in England, company number 785998, of Houndmills, Basingstoke, Hampshire RG21 6XS.

Palgrave Macmillan in the US is a division of St Martin's Press LLC, 175 Fifth Avenue, New York, NY 10010.

Palgrave Macmillan is the global academic imprint of the above companies and has companies and representatives throughout the world.

Palgrave® and Macmillan® are registered trademarks in the United States, the United Kingdom, Europe and other countries

ISBN 978-0-230-25278-3   hardback

This book is printed on paper suitable for recycling and made from fully managed and sustained forest sources. Logging, pulping and manufacturing processes are expected to conform to the environmental regulations of the country of origin.

A catalogue record for this book is available from the British Library.

A catalogue record for this book is available from the Library of Congress.

10   9   8   7   6   5   4   3   2   1
20   19   18   17   16   15   14   13   12   11

Contents

List of Abbreviations

AAA	Accra Agenda for Action
ACP	African, Caribbean and Pacific
AfDB	African Development Bank
AGRA	Alliance for a Green Revolution in Africa
APROCA	Association des Producteurs de Coton Africains
BCI	Better Cotton Initiative
BOFA	Busangwa Organic Farming Association
BSR	Business Social Responsibility
C4	Cotton Four
CCM	Chama Cha Mapinduzi
CDA	Cotton Development Assistance
CDI	Cotton Development International
CDF	Cotton Development Fund
CFA	Communauté financière d'Afrique
cfr	cost and freight
CHF	Swiss francs
cif	cost, insurance and freight
CIGI	Centre for International Governance Innovation
CMIA	Cotton Made in Africa
CSDI	Centre for Sustainable Development Initiatives
CSR	Corporate Social Responsibility
DSB	Dispute Settlement Body
EAC	East African Community
EAOS	East African Organic Standard
ECA	Economic Commission for Africa
ECOSOC	Economic and Social Council
ECOWAS	Economic Community of West African States
ENDA Diapol	Environnement et Développement du Tiers Monde Prospectives Dialogues Politiques
EPOPA	Export Promotion of Organic Products from Africa
ESRF	Economic and Social Research Foundation
FDI	Foreign Direct Investment
FLO	Fairtrade Labelling Organizations International
FNPC	Fédération Nationale des Producteurs de Coton du Sénégal
fob	free on board price

GATT	General Agreement on Tariffs and Trade
GSP	Generalized System of Preferences
HIPC	Heavily Indebted Poor Countries
ICAC	International Cotton Advisory Committee
ICA	International Commodity Agreements
ICTSD	International Centre for Trade and Sustainable Development
IFOAM	International Federation of Organic Agriculture Movements
IFAD	International Fund for Agricultural Development
IMF	International Monetary Fund
IMO	Institute for Marketecology
IPE	International Political Economy
ITO	International Trade Organization
MFA	Multi Fibre Arrangement
MDGs	Millennium Development Goals
MDRI	Multilateral Debt Relief Initiative
MKUKUTA	National Strategy for Growth and Reduction of Poverty
NAM	Non-Aligned Movement
NAMA	Non-Agricultural Market Access
NIEO	New International Economic Order
NTB	Non-tariff barrier
NSI	The North-South Institute
NSMD	Non-state Market Driven
OAU	Organization for African Unity
ODA	Official Development Assistance
ODI	Overseas Development Institute
OECD	Organization for Economic Cooperation and Development
OECD-DAC	Organization for Economic Cooperation and Development – Development Assistance Committee
PPA	Participatory Poverty Assessment
PRGF	Poverty Reduction and Growth Facility
REPOA	Research on Poverty Alleviation
SACCOS	Savings and Credit Cooperative Societies
SIDA	Swedish International Development Agency
SODEFITEX	Société de Développement et des Fibres Textiles
TanCert	Tanzania Organic Certification Association
TACOGA	Tanzania Cotton Growers Association
TCA	Tanzania Cotton Association
TCB	Tanzania Cotton Board

TIC	Tanzania Investment Centre
TOAM	Tanzania Organic Agriculture Movement
TRIMs	Trade-Related Investment Measures
TRIPS	Trade-Related Aspects of Intellectual Property Rights
TZS	Tanzanian Shilling
UEMOA	Union Economique et Monétaire Ouest Africaine
UNCTAD	United Nations Conference on Trade and Development
UNDESA	United Nations Department of Economic and Social Affairs
UNDP	United Nations Development Programme
UNEP	United Nations Environment Programme
UNIDO	United Nations Industrial Development Organization
UNMP	United Nations Millennium Project
UNU-WIDER	United Nations University – World Institute for Development Economics Research
URT	United Republic of Tanzania
USADF	United States African Development Fund
USDA	United States Department of Agriculture
USAID	United States Agency for International Development
US-GPC	Union de Secteur des Groupements de Producteurs de Coton de Kédougou
USTR	United States Trade Representative
VAT	value added tax
WCGA	Western Cotton Growing Area
WWF	World Wildlife Fund
WTO	World Trade Organization

Acknowledgements

I am grateful to many individuals and institutions for helping me to build my research on cotton into this book. William D. Coleman, CIGI Chair in Globalization and Public Policy at the Balsillie School of International Affairs, University of Waterloo, Canada, provided an incredibly high level of intellectual and material support. I am hugely indebted to him for these efforts and for his academic guidance more generally. Tony Porter and Robert O'Brien also provided invaluable insights on the direction of the argument over the years and helped move this work towards completion. I also owe particular thanks to Timothy M. Shaw for evaluating my dissertation project and encouraging its further development. Ann Weston and Roy Culpeper offered crucial feedback on my early project ideas, and implored me to articulate an argument that would resonate within academic and policymaking circles.

Many others helped me to understand parts of the story through formal interviews or informal conversations. I am particularly indebted to Daniel Drache, Brian Cooksey, Gerald K. Helleiner, Sam Wangwe, Bill Morton, Sunday Khan, Oswald Mashindano, Mwatima Juma, Joe Kabissa, Eric Hazard, Sally Baden, Abdoulaye Dia, Moussa Sabaly, Amdiatou Diallo, Alexis Anouan, Barry Alimou and Boubacar Kamissoko for sharing their perspectives and analyses with me. I am especially thankful for the time several dozen individual cotton producers took to tell me about their daily lives through interpreters, and for Pendo Kundya's subsequent translation and transcription.

I am also grateful to a number of institutions and individuals for supporting this project. The Social Sciences and Humanities Research Council of Canada afforded me with a Canada Graduate Scholarship, and the Canada Research Chair in Global Governance and Public Policy enabled the bulk of my field research. At the Institute on Globalization and the Human Condition, Sara Mayo provided exceptional organizational support and Nancy Johnson offered editorial advice on several related pieces. My research benefited materially from the Institute's Globalization Research Scholarship and also from the Globalization Internship award. The latter, taken up at The North-South Institute, facilitated the connections necessary for me to affiliate with the Economic and Social Research Foundation in Dar es Salaam and with ENDA Diapol (Environnement et Développement du Tiers Monde Prospectives Dialogues Politiques) in Dakar. Donald

Max of CopCot Cotton Trading and Niranjan Pattni of bioRe Tanzania made my interviews with cotton producers possible. Baraka Shelukindo and Keba Faty were both exceptional hosts and cultural guides. Thanks are especially due to my fellow travellers in Dar es Salaam, Liam Kavanagh and Sakari Saaritsa, and to my colleagues Jeff Ballinger, Robert Huish, Matias Margulis and Namwaka Omari-Mwaikinda.

In his capacity as the general editor of the IPE series, Timothy M. Shaw was highly supportive of this project from the outset. At Palgrave Macmillan, Alexandra Webster, Christina Brian and Renée Takken helped me to pull the pieces together. Nancy Johnson did fantastic work reformatting the manuscript and checking my references, and my new colleagues at the University of Guelph offered a high level of additional help and support. Final thanks are reserved for my long-suffering parents and for my partner Lauren Scannell. A budding anthropologist and political economic geographer, Lauren encouraged me to not lose sight of the people that depend on cotton. This book is dedicated to her.

1
Introduction: Cotton-Picking Problems Beyond the WTO

When the WTO's Cancún Ministerial meeting collapsed on 14 September 2003 I was especially intrigued. As I watched the updates and continued to piece together my thoughts I wondered how the choice that developing countries had made to pull the plug on the meeting related to my research interests. I had been thinking about the history of the collective attempts that the 'South' or 'Third World' countries had made within the United Nations system and multilateral trade negotiations to extend the benefits of the post-war international economic order to all states. As I saw it, these efforts had challenged the rich countries that had designed and maintained the post-war order to more fully embrace the ideological compromise that underpinned it.

At the risk of glossing the subtleties, this grand bargain – termed 'embedded liberalism' by John Ruggie (1982) – had been struck to enable individual governments to enjoy the policy space or autonomy necessary to pursue activist social policies. It entailed the construction of a distinctly *non*-liberal' or restrictive financial order and the simultaneous pursuit of negotiations that aimed to foster trade liberalization (Helleiner 1994: 4). From the outset, cash-poor governments strongly urged industrialized countries to broaden their discussions on trade to include issues of interest to the South. In their view, the bargain itself was too narrowly conceived. It did not address the asymmetric structure of world trade, and it did not seek to foster a redistribution of incomes and wealth to less developed countries (Steffek 2006). However, from the early 1970s rich countries gradually embraced financial liberalization and failed to rein in cross-border capital movements. As such, the compromise was in decline as developing countries issued increasingly strident calls to extend it and establish a new international economic order. After global finance was enabled, market fundamentalism became the new mantra and the interests and economic priorities of governments across the Third World diverged. As the phenomenon Sylvia Ostry (2000) has dubbed 'Ronald Thatcherism' took hold, interstate activism for a broader compromise seemed a spent force.

With this background knowledge and analysis fresh in my mind I wondered how the events at Cancún could be explained. Were the seemingly unified voices and dramatic exit evidence of resurgence? Was this instance of collective action a watershed, an outlier or something entirely new? What roles, if any, had non-state actors played behind the scenes? I pondered these questions over the subsequent days as the principal rationales behind the walkout came to light. Of these, a lack of perceived progress on agricultural trade liberalization in general and on cotton in particular seemed to me at the time to be exceptionally compelling. My interest in the latter motivating factor was cemented the following week after I had an extended discussion with a young Ugandan whose family farmed cotton. This engagement stirred my interest in the micro challenges and opportunities associated with cotton cultivation and marketing, and I subsequently resolved to pursue doctoral studies on the topic of cotton and poverty in Africa.

The ensuing two-year process of research and reflection to inform and arrive at an actionable question, methodology and research plan entailed a steep learning curve and a lot of critical thinking. Some might consider my approach to learning about how Africa's cotton problems could be framed over those years to be symptomatic of the general 'decline of deference' to elites within and beyond the academy (Nevitte 1996; Drache 2008: 5). They would not be wrong. I was willing to question the authority of high-level economists, government figures and new non-governmental elites. Many influential voices had publicly asserted that liberalization of the world cotton trade would necessarily improve the lot of the tens of millions of people in Africa who relied upon cotton production. I learned more about the policies necessary for countries to effect the transformation from dependence upon primary product exports to a more diverse, knowledge-based economy. And as I became cognizant of the downward pressures that new trade agreements variously placed on the policy space poor countries had to add more value to their exports and diversify their economies, my dissatisfaction with narrow perspectives on the 'benefits' of trade liberalization grew further still. I wondered if the long-term costs of reliance upon increased exports of cotton lint would outweigh the short-term gains from trade liberalization. In a world where the static assumptions underlying free trade theory – perfect competition, factor mobility, constant technology and zero externalities

– seemed fantastical, the prominence of the mono-causal analysis was especially troubling. I delved into the heretical literature on trade and gained knowledge of the coordination failures, information externalities, scale economies and human capital deficiencies poor countries had to overcome. I investigated the tools that they once commanded and currently have at their disposal to capture value and diversify what they produce and export (Thrasher & Gallagher 2008).

While pursuing this course of study several noteworthy economists publicly and politically opined on their preferred ways and means to end poverty and facilitate development. Much to my chagrin, the debate that ensued between Jeffrey Sachs (2005) and William Easterly (2006) more or less excluded or obscured the policy space imperative. Sachs and celebrity campaigners such as Bono pushed for increased aid flows, more debt relief and quicker trade liberalization, a perspective that equated the scaling up of resources with poverty reduction. Easterly rejected the big money solution, and painted it as highly utopian. He claimed that this type of idealistic thinking had been at the root of the aid enterprise's apparent ineffectiveness for decades, and he prescribed market-based solutions and more market-friendly policies at the country level.

As these two prospective shepherds vied for control of the same flock the rhetorical volume became so intense that another mainstream economist joined the fray to propose an ostensible middle road. Paul Collier (2007) advocated global measures to tackle civil conflict, end the adverse social outcomes associated with natural resource extraction, reduce the costs borne by landlocked economies and put a halt to the 'failed state' phenomenon. In his analysis, a focus on these issues was needed to ameliorate the conditions of life for the 'bottom billion' trapped in the poorest countries. This economist's attempt to extrapolate political prescriptions from his previous quantitative studies was nonetheless peppered with numerous polemics. These departures generally targeted people who had dared to question the free trade faith. In my reading, his presentation also obfuscated the policy space issue.

I became convinced that the implicit departures that these three thought leaders had made from their roots as avowed social 'scientists' into the realm of polemical political economy had, to borrow Easterly's provocative phrase, done a 'little good' and maybe even a little 'ill'. While their contributions certainly informed discourses on development and generated public interest in poverty, they pulled the wool over the eyes of the socially concerned by unduly limiting debate. Their outputs 'disappeared' a canon of heterodox thinking on economic development.

I concluded that to get as close as possible to understanding the cotton-poverty nexus I would have to remove the woollen blinders and hope that when the product was ready for market, others were willing at least to remove any cotton that had been stuffed into their ears long enough to hear me out.

Beyond my preoccupation with this somewhat false debate, the new force on display in the lead up to the July 2005 Live 8 concerts and the Group of 8 meeting at Gleneagles drew my attention to questions of authority and power and helped me to secure my central research question. John Kenneth Galbraith (1967: 262) once painstakingly detailed a growing blurring between private and public authority structures. As transplanetary connections among people subsequently increased and occasionally took 'supraterritorial' forms – the essence of Jan Aart Scholte's (2005: 59–60) textbook definition of globalization – scholars that were attuned to these shifts recognized a need to extend Galbraith's analysis to the global level. They attempted to situate incipient types of 'private' authority and the relations of this authority to legitimate public power (Cutler et al. 2001). Watching from the sidelines in 2005 I marvelled at the aspects of the phenomenon my teacher Daniel Drache subsequently coined the 'unprecedented reach of the global citizen' that were on display at Gleneagles. Still, I worried that there was little public debate or even publicly available knowledge of the efficacy of the new non-governmental actors *vis-à-vis* poverty reduction and economic development. The attempts of 'private' authorities and individual corporations to embrace social responsibility more generally were similarly opaque.

In hindsight, my resolve that summer to concentrate on constructing an evaluation of these new global factors as regards cotton can be viewed as an attempt to transcend what Arjun Appadurai (2000) referred to as an embryonic 'double apartheid' linked with globalization and globalization studies. He identified a growing divorce between debates of a 'parochial quality' on globalization within the Ivory Tower and popular discourses on how to ensure cultural autonomy and economic survival worldwide. Appadurai argued that the second face of apartheid was the conspicuous inability of the poor the world over to engage in national and global dialogues about globalization. The nasty corollary of their silences or exclusion was clear enough: the failure of authorities at all levels to often act upon the concerns of the poor effectually or even at all.

Research problem and questions

Accordingly, I turned my lens outwards and set out to detail to the best of my abilities the totality of relationships that have impoverished

Africa's cotton producers down to the present. I wanted to understand and re-present the ideas that policy elites and the poor themselves thought necessary to alleviate, reduce and eradicate poverty. Through this situational analysis I aimed to produce a baseline for a broader evaluation of the impact of globalization than was evident in the literature. The need for a more wide-ranging analysis was brought home over two years later when an edited collection – *Hanging by a Thread: Cotton, Globalization and Poverty in Africa* – was released (Moseley & Gray 2008). Taken together, the comprehensive analyses contained within this impressive collection revealed just how idealistic it would be to believe that liberalization alone would eradicate poverty problems. However, beyond the thorough review of the prospects and pitfalls of organic cotton offered by Brian Dowd, the other contributors said very little about the emerging forms of private and non-governmental authority operative at the global level and across the continent whose actions were increasingly consequential. Oxfam was mentioned only fleetingly, and several other entities and initiatives that aimed to inform governments or directly govern African cotton – including the Geneva-based IDEAS Centre, the Better Cotton Initiative (BCI) and the Cotton Made in Africa (CMIA) label – were simply not discussed.

As mine was to be a study grounded in and informed by the budding field of international political economy (IPE) and the distinctive Canadian political economy tradition (Watkins 2003), it was possible to select an original and very large question to drive my research forward. I asked *what impacts, if any, new governance trends such as nongovernmental policy advocacy and approaches to corporate social responsibility (CSR), were having on the historic relationships between cotton production and poverty south of the Sahara.* My advisory committee signed off on this difficult problem knowing that I would have to construct a baseline analysis that would rely upon qualitative and ultimately fallible data. At the outset I had alerted them to several rationales for the proposed breadth. For one, it appeared to me that there was a need to build an accessible, decolonized and multidimensional understanding of poverty and the factors that have maintained it. I believed that this approach could counterbalance the income-centric understandings of poverty that advocacy campaign materials, the big-name economists and popular discourses on Africa's cotton problems had fostered. Notwithstanding the reality that my findings on the relationships between cotton and poverty would be ultimately contestable – poverty remains an essentially contested concept – and the fact that I was setting out to evaluate the impacts of new and intensely dynamic phenomena, the project moved forward. I had eschewed pretensions to social 'science' but resolved to the best of my abilities to produce a story that rang as true as possible to what my friend Robert Huish of Dalhousie University has termed the 'sounds from the ground'. As long as this tale could draw attention to the ways that globalization was enabling people to overcome adversities or to the aspects of it that were entrenching unequal outcomes I figured that it would at least be relevant for the people and institutions directly impacted by or involved with Africa's cotton 'conundrum' – the growing reliance of Africans upon exports of raw cotton. If things really worked out and luck was with me, I hoped to also contribute in some small measure to at least a few debates of varying importance within the political economy, development policy and globalization literatures. However, the overriding consideration was to simply apply insights from these fields and other non-academic sources in a manner that helped to produce a robust explanation.

That being said, none of the existing studies of African cotton have been rooted in an IPE approach. Social historians (Isaacman & Roberts 1995), development economists (Dercon 1993; Hazard 2005) and economic geographers (Moseley & Gray 2008) have produced the principal edited collections on the subject. Policy-oriented professional researchers have compiled the biggest institutional publication on the topic (OECD 2006) and historians have authored the dedicated book-length monographs on country-level conditions (Isaacman 1996; Bassett 2001). Economists whose work has drawn heavily upon the new institutional economics (Poulton et al. 2004a; 2004b; Tchirley et al. 2010) and sociologists working to develop global value chain analysis (Gibbon & Ponte 2005) have been behind major collaborative research endeavours that have sought to analyse the impact of competition on the sub-sector and chart governance trends that bear upon Africa's broader basket of agricultural commodity exports. All of these studies have shed light on the factors that have impeded the livelihoods of cotton producers. Unfortunately, they have tended to concentrate on the micro and domestic levels of analysis, or focused attention only on how particular international institutions or the changing demands of downstream buyers have affected the African cotton scene. While researchers in this area have not yet attempted to appraise the overall efficacy of the new global phenomena, several have conducted publicly unavailable external evaluations of the development effectiveness of important non-state actors such as Oxfam International and the Geneva-based IDEAS Centre.

I attempt in this book to offer an encompassing global analysis of the relationships between cotton production and poverty and how these might be changing in the present era of globalization. I argue that cotton and poverty are historically and empirically linked, and that the impoverishing reality of cotton production in any given African locale cannot be attributed exclusively to economic structures, or to power relations or cultural norms that have prevailed within households or communities, or at the domestic or international levels. While particular factors operative in specific cotton zones and in households within those zones are to a certain extent indeterminate and variable, I found that factors associated with globalization have changed the framework for poverty reduction where they are operative. Globalization has increased the likelihood that rising numbers of cotton farmers will endure fewer impoverishing relationships and as a consequence – to borrow a line from Amartya Sen (1999) – be capable of leading lives that they value more. Even so, progress on the range of factors that hinder wellbeing across the cotton-producing countryside is not uniform. There is little in the tale that I elaborate below to suggest that a new age of poverty-free cotton farming is upon us.

Organization of the argument

To make this case, Chapter 2 commences with an elaboration of the connections between cotton production and poverty that were evident during the colonial and neo-colonial eras. It moves on to discuss how poverty outcomes shifted in Tanzania after the government pursued agricultural adjustment and ostensibly 'liberalized' the cotton subsector. Drawing upon my case study research in Tanzania it then offers a brief sketch of the human face of poverty in that country today. The next chapter discusses global governance of the cotton trade, or the lack thereof, from the post-war era down to the present. It recounts the failed attempt of cotton-producing and consuming countries to build an international cotton agreement and establish an institution known as the Cotton Development International. Chapter 3 also describes the attempts that have been made to govern cotton at the WTO, and it argues that even if progress on the cotton file in Geneva were to be achieved many poverty-maintaining factors would be left intact.

Chapter 4 presents the findings from my original research on cotton and poverty in Tanzania. Here I articulate a basket of specific reforms that aim to overcome poverty and what I believe to be the principal dynamic factors that constitute the limits of the possible as regards poverty reduction. In my view, the latter structural and policy factors include governance reform and resource mobilization, development assistance, biotechnology policy, foreign direct investment and lint export dependence. As I see it, each of these factors can have poverty-maintaining or poverty-reducing effects. For example, it cannot be said with any degree of scientific certainty or precision that the provision of more grants or concessional loans will necessarily reduce poverty, or that a welcoming stance *vis-à-vis* foreign investors is reliably pro-poor or systematically poverty inducing. To sketch the state of play as regards these broad categories I draw upon insights from the field. This discussion enables me to subsequently evaluate how the new instances of global governance – NGO advocacy and CSR codes – variously take advantage of these factors or remain subject to the constraints that they can pose.

Since a pan-African baseline scenario for the evaluation of the emerging global governance phenomena was unavailable and I did not command the resources necessary to construct one, I had to make use of a second-best proxy. My Tanzanian findings and my knowledge of other diverse needs elsewhere became the benchmarks for evaluating the various impacts of NGOs and CSR codes on poverty. The results of this exercise are presented in Chapter 5 and Chapter 6. In the former, I make use of Jan Aart Scholte's list of the costs and benefits associated with civil society engagements at the global level and my own baseline understanding to evaluate and compare the impacts that the IDEAS Centre, Oxfam International and other non-governmental organizations operative in Tanzania have had on poverty. I find that many non-state actors have embraced a relatively weak 'norm' for the amelioration of the cotton problem, and that North-South divides and a lack of pan-Africanism have plagued efforts to enshrine it. In spite of this, I argue that there are reasons to be hopeful moving forward.

In Chapter 6, I outline the state of the debate on corporate social responsibility. Here, the Better Cotton Initiative, the Cotton Made in Africa approach and the bioRe certified organic project in Meatu, Tanzania are each subjected to critical scrutiny. Of these, bioRe stands out as an exemplar of the poverty-reducing potential of CSR. As I argue towards the end of the chapter, however, the prevailing climate of social irresponsibility amongst conventional buyers and the other global initiatives constitute real and potential threats to the bioRe project's durability and replicability, and to similar projects that rely upon 'hardcore' social responsibility methods such as third-party certification. Finally, in the reflections and conclusions, I ponder my international political economy approach and the thoughts that drove the work forward. The findings are then drawn together and their implications for theory and for policy are discussed. I identify a need for scholars to think more concretely about the possibility that competition to set standards for social responsibility could have unintended and deleterious consequences.

In policy entrepreneurial fashion, I also float the idea that if the United States of Africa were to take flight, cotton might be an interesting target to test the potential for regional integration *and* CSR to reduce poverty and enable Africans to capture more opportunities to add value to cotton. A pan-African effort to coordinate cotton systems that attempted to replicate the levels of collaboration and cooperation that were evident between state actors and the new global 'governors' during the successful development of the East African Organic Standard would be a consequential governance experiment. Though such an attempt to generalize the benefits of CSR could be associated with numerous benefits and costs at all levels, my research clearly indicates that the social, economic and environmental costs currently borne by conventional cotton farmers are disproportionately high, and that there are strong rationales for the pursuit of a more harmonized approach to CSR that is more clearly aligned with the imperative of poverty reduction. A pan-African approach could constitute one potential way to transcend the coordination failures of current CSR efforts. It is by no means certain that African governments are capable of heeding this unorthodox advice or that they would be willing to collaborate with the emerging global governors of African cotton. However, the prospect that higher and sustained levels of greener growth and poverty reduction could flow from such interactions is backed by the research presented below.

Research methodology

This cross-cultural study proceeded inductively. Preliminary hypotheses and conclusions were developed in a bottom-up fashion, and the case study research relied upon qualitative methods. I initially selected Tanzania and Sénégal for case study analysis specifically so that I could compare the impacts that the new global phenomena were having under diverse conditions. I felt that the evident cultural, geographic, historic and linguistic differences between these countries, and the divergent reform paths that they had embarked upon, were ripe for comparison. While these states were not amongst the leading lint producers or exporters in Africa, cotton was often one of the top foreign exchange earning agricultural exports of both countries. Instead of choosing to go where cotton was a national preoccupation or where it garnered little attention, I chose to do field research in countries that were dependent upon cotton, but not inordinately so. Tanzania and Sénégal were also logical choices as the non-governmental and new corporate approaches that I was interested in understanding were clearly evident. I planned to conduct elite, semi-structured and in-depth interviews in both countries, and to retrieve relevant primary documents while based in Dar es Salaam and in Dakar over a six-month period. Additionally, I aimed to interview direct producers who had engaged with non-state actors or participated in new CSR systems, and also to talk to those who had been more isolated and less exposed to new players or approaches. In particular, I wanted to pursue field investigations of the bioRe project and a conventional buying operation in Tanzania. I also hoped to engage in French with officials at Sodefitex, the Sénégalese cotton company, to learn about an embryonic certified fair trade project near Kédougou.

To facilitate my field research I established institutional affiliations with ENDA Diapol, a Dakar-based think-tank, and also with the Economic and Social Research Foundation (ESRF) in Dar es Salaam. Prior to commencing my fieldwork, Eric Hazard, now Oxfam International's regional economic justice campaign manager for West Africa, and staff members at the ESRF, conducted ethical reviews of my proposal. After these individuals cleared my work the McMaster University Research Ethics Board granted me ethical approval to proceed with the study. As a result of this rigorous research ethics process I was able to design culturally appropriate strategies to obtain informed consent. I also became versed in social risk minimization and developed contingency plans to ensure that my data was kept confidential and secure. Prior to my departure for Dar es Salaam I also applied to the Tanzania Commission for Science and Technology for the necessary research permit and received this official document upon arrival.

My interviews at the country-level were incredibly informative and I realized the desired 'snowball' effect. I also collected a considerable amount of primary data from both countries that would have otherwise been unavailable. However, after I commenced the write up it rapidly became clear that due to space and time constraints I would have to scale back the ambitions I had at the outset to complete a comprehensive comparative analysis of Tanzania and Sénégal. Consequently, a situational analysis of poverty and cotton in Sénégal was not included in the initial manuscript or in this book. Instead of drafting this second set of poverty benchmarks, I drew upon the content of my Sénégalese interviews and the French-language primary documents I had retrieved to speak to the problems and progress of non-governmental policy advocacy at the global level, and also to the development effectiveness of organic and fair trade systems. The comparative component of my case study approach consequently shifted, and I focused intently on the relative strengths and weaknesses of conventional cotton buying operations and certified organic operations in Tanzania.

On a cautionary note, the following pages attempt to share knowledge about a complicated, value-laden and politically controversial topic. I do not pretend to have arrived at all the answers about the emerging global governors of African cotton. That said, I do have a story to tell and I hope that it is as respectful a tale as can possibly be told on the subject. While I am at times highly critical of particular institutions, ideas and power relations, the aim all along has been to understand the international political economy of Africa's cotton conundrum in order to evaluate it and foster change. It might also be true that the multidimensional understanding of poverty that I have embraced broadens the concept so much that it is difficult to pinpoint where it ceases to apply and where other concepts such as inequality begin. Be that as it may, I have attempted to explicate the multifarious factors that impoverish people and communities with a view towards producing a robust analysis that is accessible to a broader audience and relevant in its own small way to the global effort to make cotton work better for the Africans that grow it.

Whatever the specific or context-dependent impediments that individual producers and households across the continent have faced, it is clear to me that many have been up against it (see Appendix A). And they often still are. Take for example the wild claims espoused by one conventional buyer that I encountered:

> We give cotton farmers a lot of money. I don't know where they are putting their money. They've got a lot of money. They make money. They don't know how to manage their money. It's a money management problem and that is all.

When I heard these beliefs I offered no retort. I did not stop to tell him that the price per pound of lint reached $12 in real terms during the US civil war and that it was currently averaging roughly 60 cents (*Financial Times* 2007). I failed to offer up the 'fact' that 95,000 ultra-high-net-worth individuals around the globe commanded about 35 per cent of the world's financial wealth (Thal Larsen 2007). Nor did I ask him what he thought about the idea that rapid oil price rises had cancelled out the sum total of debt relief that had been extended to his oil-importing country over the previous three years and the new aid flows that had been disbursed to his country over the same period (Crooks & Green 2007). I resisted the urge to blurt that a *Financial Times*/Harris poll had found that globalization generally and corporate leaders specifically were held in low esteem across the rich world *before* the credit crunch of 2008, and refrained from asking his thoughts on these matters (Giles 2007). I simply recorded his views and moved on hoping that

I would be able to recount them in at least a somewhat consequential way.

As this book was under construction the financial crisis took hold. Early on it appeared that inflows of capital to Africa were set to dry up and that capital outflows would increase. The dark possibilities that African governments would be crowded out of bond markets and receive fewer aid dollars also looked nearer to reality. China – Africa's emerging financier and the number two importer of the continent's raw materials – seemed set to fill these voids, but the anti-poverty payoffs of their self-interested engagements were seen to be ambiguous at best (Taylor 2006). Simultaneously, rhetoric on the need to address the global food price crisis outpaced the uptake of new initiatives to enable the poor to be more food secure or enjoy food sovereignty.

In this context, Africa's reliance on the production of inedible export crops like cotton appeared to be a recipe for hunger and immiseration. Clothing sales in the United States had plummeted with the onset of the crisis and seemed poised to fall even farther or at least stagnate. Moreover, down to June 2010 the United States had not revised the aspects of its cotton subsidy system that had fuelled the production of a domestic surplus and the dumping of that excess supply on to the world market (Politi & Wheatley 2010a; *Financial Times* 2010). This source of considerable downward pressure on the world cotton price continued unabated, reducing the incomes Africa's cotton exporters could earn from the trade and lowering their ability to pay farmers higher prices at the farm gate.

Despite the gloom, the new governance trends analysed below have persevered. Worries that corporate social responsibility would start to appear faddish in an era of tight credit, corporate retrenchment and consumer cost-cutting have not materialized. Similarly, non-governmental organizations have continued to be advocates for trade policy reform and poverty reduction more generally. While it is now all too clear that market failures and governance failures in the areas of global finance and world trade can have serious consequences for the poor in Africa, this book shows that the new governors of African cotton have actively pursued significant interventions. While some NGO actions and CSR initiatives have demonstrably fostered poverty maintenance, others have cultivated sustained poverty reduction. This study will hopefully stimulate those interested in knowing more about impact of global governance on the prospects for poverty reduction, and inform those interested in ending the disempowerment and oppression historically associated with cotton production south of the Sahara.

2
Historic Relationships Between Cotton and Poverty

This chapter discusses historic relationships between cotton production and poverty south of the Sahara from the colonial era through the 1980s, and stresses the enduring nature of these linkages. To start, an historical narrative of cotton and colonialism is offered that highlights the ideas, institutions and power dynamics that characterized the era of Sub-Saharan Africa's incorporation into metropolitan spheres as a source of low cost material inputs for European textile mills. Subsequently, the colonial legacy of state involvement in cotton markets evident during the post-independence period is detailed. This section focuses on the ways these systems precluded the betterment of conditions of life in many cotton-growing zones, and the subtle strategies producers deployed to resist the aspects of cotton cultivation that they perceived to be exploitative. An account of cotton sub-sector reforms in Tanzania that generated income poverty, fuelled a polarization of incomes from cotton and skewed the distribution of opportunities cotton farmers had to reap gains from their crop is then presented.[1] A concluding section draws insights from the three eras together and presents a brief composite example of the impoverishing obstacles an average cotton grower might face going about their life in the present day. It underscores the point that attempts to appraise the anti-poverty potential of cotton relative to the production of other crops that draw solely upon anecdotal or micro-evidence of correlations between the choice to grow cotton and higher smallholder incomes are wrong-headed. Robust evaluations of the cotton-poverty nexus require attention to a profusion of other environmental, ideological and institutional factors, and to power relations within and beyond rural households.

Cotton, colonialism and poverty

Prior to the era of European imperialism and formal colonization, cotton played a minor role across the diverse societies of Sub-Saharan Africa. A particularly strong handicraft textile industry was to be found in the West, though no communal fields or compounds were devoted entirely to the production of cotton. Social historians have attributed the minor role of the crop to the region's inhospitable climate. Contributors to a collection edited by Allen Isaacman and Richard Roberts (1995) have shown that precipitation patterns were also much more amenable to successful cotton production in Algeria, Egypt and Morocco than they were south of the Sahara. For cotton to succeed in the absence of irrigation technologies regular rains must fall over five consecutive months and let up shortly before the harvest. Precipitation levels across the equatorial zone were variable and unpredictable, and as a consequence, held back experimentation with cotton and discouraged its widespread adoption. The region's high level of biological diversity also worked against the scaling up of production. An incredible array of insects, fungi and weeds posed significant threats to the harvest and necessitated the constant attention of growers. High average daily temperatures also had a notable effect on production and consumption. Amongst other factors such as permissive and practical social norms, generally hot conditions reduced the necessity and demand for the everyday use of textiles as clothing. As a result, the range of potential final uses for any harvested cotton was more constrained than it was in North Africa.

According to Isaacman and Roberts, soil compositions in the south were also disadvantageous. Across tropical Africa soils were often deficient in nutrients relative to those found in the North. In particular, they were more prone to leaching and also to erosion. Africans that did choose to grow cotton were well aware of the potential pitfalls of their choice. It is likely that cotton was typically intercropped with nitrogen-fixing food crops such as yams or legumes to minimize its impact on the weak soils, and maximize its chances of germination and survival. Producers also knew that intensive efforts were required from the time their crop took root until the bolls opened and were ready for picking. Whether for the needs of the compound, community or the handicraft trade, cotton cultivation required strong skills and good fortune. Prior to imperial contact, cotton remained for the most part a peripheral rural occupation of limited social and economic importance.

When Europeans came to Africa in the sixteenth century they did not come to acquire cotton via the barrel of a gun. Europeans discovered that their trading firms could exploit the evident local demand for handicraft textiles for practical, ornamental and ceremonial uses only after two centuries of intensifying contact, plunder and unequal exchange relationships. In England, the belief in the new African market was fuelled by the progressive application of the division of labour and the invention and deployment of productivity-enhancing technologies that reduced unit costs and increased cloth production volumes. By the 1800s the relatively undifferentiated textiles produced in Lancashire's mills were seemingly tradable across wide swaths of tropical Africa. The subsequent creation of this outlet for increasing cloth output averted a demand crisis and added another channel to England's income stream. As with the savings derived from trade in Africa's precious metals, gemstones, slaves, timber and agricultural produce, savings from Lancashire's cloth exports were re-invested in England's highly protected industrial development. Further industrial innovations in Lancashire fuelled evermore increases in textile production down to the 1860s, and a consistent stream of those goods flowed to Africa to compete with homespun cotton products.

During the first half of that decade it became relatively more difficult for the British to secure adequate supplies of cotton lint. Lancashire industrialists recognized that they had to address this new imperative. Expanded production capacities in France, Germany and Portugal had increased pressure on international traders to deliver ever-greater volumes of lint. These state-led developments led to significant rises in the world lint price. The US Civil War and a boll weevil epidemic also reduced the amount of lint available for traders to source and raised the price further still. This so-called cotton 'famine' was the first raw material supply crisis of the industrial era (Farnie & Jeremy 2004). To reduce their lint procurement costs enduringly, and enhance the growth and competitiveness of their new textile and clothing industries, British and European manufacturers pressed their governments to promote cotton production abroad.

These efforts gained speed subsequent to the Berlin Conference of 1884–85 that established colonial boundaries, formalized the process of rent extraction and gave European powers impetus to assert absolute control over the continent. In the aftermath of the Conference, metropolitan governments sought to ensure that returns on investments and trade with their colonies were higher than they would have been in the absence of their colonial states, and more lasting than those of their European rivals. As regards cotton, colonial states were tasked with guaranteeing that additional volumes of low cost lint reliably flowed to their respective metropoles.

From the outset unstable market conditions and environmental challenges reinforced the significance of these directives. During the 1890s
the lint price fluctuated and then rose considerably as supplies came
under pressure yet again. Growers and ginners in the south of the United
States realized new productive efficiencies during that decade and raised
their outputs. However, after exports of textiles and other US manufactures began to surge around the mid-1890s, domestic textile production consumed the new supplies and ate into the volume of lint available
for export (Irwin 2003). Severe pest outbreaks in several cotton zones also
led to massive and unpredicted production declines. The speculative
activities of New York traders compounded lint price rises and variability
(Isaacman 1996). While European firms essentially had access to a captive
lint supply from the colonies, Africa had not previously been a consistent
source of competitively priced lint. As the second supply crunch in as
many generations took hold, Europeans warmed to the prospect that
African procurement could be made viable and enriching. Members
of a number of new organizations that aimed to arrive at long-term
solutions to the issues of lint scarcity and price instability articulated this
perspective.

Through the turn of the century and into the early 1900s, cotton-
growing associations were established in Britain, France, Germany and
Portugal (Baillaud 1903: 132). These new groups commenced a search
for a fresh approach and typically were composed of all private market
participants and relevant public officials. Steve Onyeiwu (2000: 91) has
documented how members of the British Cotton Growing Association
endorsed an extensive search for an alternative and cheaper source to
US lint supplies. Through this work the attention of King Edward VII
was drawn to the issue, and in a 1904 speech he stressed the imperative
of maximizing cotton production in the colonies (Mamdani 1996: 37).
The other national coalitions made similar moves to realize mercantilist
objectives.

A prominent assumption underlying the early twentieth century
drive to make African cotton work for European development was that
agricultural technologies could be deployed to minimize the impact of
the environmental factors that had previously impeded cotton cultivation. Colonial officials were instructed to organize production systems
scientifically to ensure the timely application of productive inputs, and
the efficient use of those inputs, to maximize annual seed cotton outputs. After initial attempts to meet these demands failed to achieve
results officials explained away their evident lack of progress. In particular, they asserted that the supposed 'character' of their rural Africans

subjects prevented the uptake of approaches that Europeans considered to be 'rational'. The view that African agriculturalists were chronically lazy was broadly shared, and many officials believed that their apparent lack of acquisitive or self-maximizing states of mind could be reversed only if they were given strong incentives to change their life ways. Debates in Europe ensued over the relative merits of the market, coercion, or the use of brute force to overcome the imagined nature of the African farmer and enhance production.

Arguments for the coercive and forceful integration of Africans into cotton-growing schemes won out in many colonial states. Production programmes in the Belgian Congo, Côte d'Ivoire, Mozambique and Oubangui-Chari, the colonial antecedent to the République Centrafricaine, relied upon significant amounts of coercion as well as the actual use of force. In Mozambique, nearly 800,000 people were compelled to formally register in the cotton programme. Isaacman (1996) has documented how colonial authorities there exerted absolute control over the cotton economy. Officials selected the lands to be used for cotton cultivation, dictated work schedules for crop husbandry and also set prices by fiat at market locations that were often many days walk from the fields where cotton was grown. Under this scheme, Mozambicans were told when and where to live and work, and did not enjoy enough freedom from their obligations even to produce food for themselves. Cotton growers were compelled to labour on the land holdings of collaborationist chiefs or big men, and these unremunerated tasks took precedence over own-use food production. As men from the south of the country continued to migrate across the frontier to the mines in Witwatersrand in search of better incomes, women bore a disproportionate share of these workloads.

Officials in colonial Mozambique perceived a need for surveillance and discipline to ensure the system's success. They understood that if registered farmers were treated leniently, many might revert to their historic preference and simply choose not to cultivate cotton intensively or extensively on the desired scale or even at all. Prior to the colonial era, each year when it came time to decide whether to sow food crops or to plant cotton, agriculturalists throughout the land thought first about filling their bellies. By way of their choice to sow edible crops that were less difficult to farm than cotton, Mozambicans demonstrated an important rational impulse. This logical, thought through process of selection contradicted the paternalist view that Africans ostensibly lacked the ability to reason. With no incentive to realize the irony of their actions, colonial functionaries set out to eradicate the possibility that farmers could go back to their old ways. The state made side-payments in the form of credit, bonuses and commissions to traditional authorities to impose the cotton economy. By offering these inducements to implement production plans and enforce rules, the government was able to avoid overstretching its relatively limited regulatory and supervisory capacities. For decades this system to co-opt local leaders ensured that forced Mozambican production and low value added cotton exports supported the growth of Portugal's textile industry.

Colonial cotton programmes all over Africa nevertheless failed to reach production targets. After an extensive search for an answer to their local shortfall, officials in Côte d'Ivoire eventually settled on an obvious resolution. They argued that seed cotton output would only increase if prices were set at a level equal to or relatively higher than those offered for other cash crops. Like the shrewdness Mozambican officials had previously detected on the farm, this late-colonial conclusion on market incentives contradicted the earlier mythology of the 'irrational' disposition of the 'African', but did not resonate widely (Bassett 2001). Officials elsewhere reasserted an older option that was potentially more amenable to cotton traders, textile firms and the colonial project. They generally argued that attempts to raise yields through the provision of productive inputs and the application of advanced agronomic techniques would avert further failures (Talbott 1990: 103). Policymakers that adhered to this view nonetheless wanted to maintain relatively or absolutely low farm gate prices, and continued to consider these prices to be the basis for colonial trade and also its *raison d'être*. Beyond Côte d'Ivoire and Uganda, debates about cotton focused increasingly on the practicalities of introducing scientific and managerial innovations. Regarding the latter, the optimal scale of production became an issue. Several colonial regimes studied, weighed their options and proceeded to test the relative merits of smallholder intercrop production and large monoculture operations.[2] However, analysis and experimentation to pinpoint the most advantageous approach to management and the application of science more generally did not lead to the resolution of underperformance issues. It is likely that the principle of lowest cost – the *de facto* founding principle of Africa's cotton economy – and the ambiguous relationship of rural Africans to cotton were significant constraints. These realities undermined modifications that were made in the name of science or scientific management and compromised the ability of these means to durably increase seed cotton outputs.

African cotton production only began to approach the volumes desired by Europe after colonial regimes that had pursued coercive systems moved to abandon direct compulsion and mitigate other egregious aspects of their policies to exhort people to grow more crops. British colonial officials had discarded their coercive programmes from the 1920s, and the French followed suit during the 1940s. Beyond the Portuguese and Belgian spheres of influence where schemes remained intact, officials opted instead to espouse policies that compelled production more indirectly. Many chose to exact taxes on cotton farmers. This tactic had been previously deployed by colonial states to coerce rural dwellers away from their pre-colonial sustenance economies into dependence upon cash cropping and participation in the money economy (Hyden 2006: 14). In effect, this moderation gave many producers an unenviable choice. They could pay their taxes through cultivating and selling their crop in marketing systems that had been designed to be explicitly low cost, or they could choose to not be able to pay their taxes at all.

Into the era of formal decolonization only Mozambique continued to rely on the direct use of force. There, the sole notable change after the Second World War was that administered farm gate prices began to be set at slightly higher levels. Despite a decline in the world lint price the state was able to grant this concession due to the fact that its system guaranteed Portuguese buyers a price that was considerably lower than those that alternative sources could offer consistently. After the price rise Mozambican growers no longer required the total proceeds from sales of 100 kilos of seed cotton to purchase a single loincloth (Isaacman 1996). This increase enabled the average woman that sold this amount to literally buy a handful of additional goods. Consequently, the knife-edge upon which these farmers continued to subsist remained unaltered. They certainly did not enjoy anything near to what has since been termed 'food security'. This scheme obligated the monocultural cultivation of cotton and did not allow for crop rotations or intercropping techniques that could have replenished soils, ensured the future productivity of the land or provided producers with a subsistence buffer if their cotton failed. The cotton harvest also coincided with the most labour intensive cultivation period of many traditional staple foods. As such, the programme endogenously generated a labour bottleneck. Growers in Mozambique faced the prospect of food crop failure while they fulfilled their duties on their supervised cotton plots or in the 'commander's field'. Mozambicans did not resign themselves to this bleak reality. Many resolved to overcome their tormentors and challenged maltreatment covertly and discreetly. In so doing, they chose resistance tactics that were similar to those upon which other cotton growers south of the Sahara drew.

During the colonial era cotton producers used modest, everyday means to confront oppression in all but one or two noteworthy and bloody instances.[3] While no armed conflict was ever fought explicitly or solely over cotton, deprivation and malcontent that stemmed from the implementation of a forced cotton production scheme set off the 1905 Maji Maji rebellion against German rule in Tanganyika. In 1902, the Governor there had decreed that each village would have to cultivate and harvest a common plot of cotton for export to the fatherland (Rodney 1972). To enforce this decree colonial authorities pulled men away from their sustenance food plots with the ostensible promise that they would share the profits with their headmen and the marketing organization. This move led to an outbreak of hunger across an otherwise fertile countryside. It also challenged the traditional gendered division of labour within households and created social turmoil. When a severe drought hit in 1905 the spectre of starvation loomed ever larger over the rural population and male workers of the Matumbi ethnic group that had been enlisted in the programme resolved to fight. On the night of 31 July, cotton growers drove their oppressors from the cotton-growing area around the Rufiji river basin (Iliffe 1972: 10). As the conflict spiralled, an influential leader and prophet, Kinjikitile Ngwale, became a medium for the spirit Kolelo of the Uluguru Mountains and took the title of Bokero. In his new incarnation, Bokero propagated the belief that the Tanganyikans had been called upon by Kolelo to stop paying their taxes and force the Germans from Africa. He encouraged his followers to drink a potion or war medicine consisting of water, castor oil and millet seeds that they believed would turn German bullets into 'maji', the Kiswahili word for water (Ocaya-Lakidi 1977: 157). During the subsequent open revolt against German rule, 200,000 to 300,000 Tanganyikans were killed, and several districts were permanently depopulated (Hull 2003: 161).

Resistance to the forced production of cotton generally took more subtle forms. Africans defied authorities by planting food crops at concealed locations in their cotton plots, or by cultivating secret food gardens. Growers developed systems of collective knowledge sharing or self-help so that word quickly spread when overseers were absent and they could work these patches covertly. They also did not simply resort to deception and subterfuge to augment their subsistence food supplies. The Isaacman-Roberts edited collection documents many instances where growers chose to limit their productivity on the job and also withheld their labour power for short periods of time. Cotton producers in Congo, Malawi, Mozambique, Oubangui-Chari and Tanganyika also sometimes chose a more overtly political resistance strategy. At times they boiled or cooked their seeds before planting them in the hope that they could later convince officials that their lands were not well suited for cotton (Isaacman et al. 1980). In colonies or particular regions where land was accessible or abundant, farmers also attempted to subvert repressive cotton systems by taking flight under the cover of darkness and rebuilding their lives elsewhere. According to Isaacman's extensive review, agriculturalists also opposed cotton through cultivating less than the officially required amounts, planting after designated dates, weeding fewer times than required, illegally intercropping cotton with food crops or allowing their cotton to rot (Isaacman 1993: 241).

Where people pursued this range of clandestine methods they demonstrated that they were not simply passive victims of colonial subjugation (Bassett 2001). Even in the totalitarian colonial society of the Belgian Congo cotton became a principal rallying point for a radical collective questioning of forced cultivation (Jewsiewicki 1980). The prevalence of creative and restrained approaches to subversion ensured that cotton was simply never cultivated on the scale that authorities demanded. These tactics lowered production volumes and undermined the prevailing climate of opinion on the necessity of compulsory cotton schemes. Lint shortfalls were also at the root of the abolishment of the last remaining cotton command economy in 1961 when Portugal's Prime Minister Salazar decreed the end of the Mozambican scheme.

Even after colonizers granted agriculturalists the formal right to choose to participate in cotton markets many farmers who continued to cultivate cotton also relied upon low-level resistance techniques to get by. The choice to grow cotton often enabled farmers to gain preferential, subsidized or credit-based access to pesticides or fertilizers. At times of need producers diverted these inputs to other uses. They applied them to their own food crops or passed them along to their families or neighbours. Prior to harvests cotton farmers also occasionally chose to side-sell these goods on illicit secondary markets to earn cash that could be used to purchase additional food or other essential items. Historical evidence suggests that the biggest and best-connected farmers reaped disproportionate benefits from this particular covert practice (Isaacman 1996). Growers also made use of a more universally available survival mechanism before delivering their crop to single-buyer or monopsony markets. According to Thomas Bassett, and to Isaacman and Roberts, many added water, stones or sand to their cotton to increase its weight and marginally raise their earnings.[4]

These acts of resistance were not solely driven by individualism. To a certain extent the interest producers had in enhancing the collective welfare of their communities also motivated their behaviour. Owing to the imposition of levies African farmers were well aware that they had a mutual need to obtain cash. Incorporation into asymmetrical colonial trading relationships also gave them a shared experience that augmented pre-existing lineage, kinship, spiritual and other social bonds even as it challenged traditional authorities and cosmologies. Cotton cultivation itself was the bitter fruit of the colonial encounter. It also gave rise to a common identity in many zones. In the West, rural people that produced cotton began to self-identify as 'cultivateurs de coton'. While producers there typically consented to the *status quo* of external domination their acquiescence was born of necessity and as noted above, did not preclude occasional disobedience. Cotton growers consistently demonstrated their reluctance to fully comply with the norms, rules or directives authorities attempted to enforce, and refused on *ad hoc* bases to be complicit in the exploitation of their communities. Resistance also exuded a community orientation insofar as the low-level tactics employed were relatively more difficult to police than large-scale sabotage, protest or violence. This approach reduced the risk of extensive and excessive retribution in contexts where farmers enjoyed few legal protections. It stretched the capacity of local comprador and colonial administrators by forcing them to address defiance on case-by-case bases. As such, these techniques were highly replicable. Farmers that had to earn an income from cotton also learned of insubordination and successfully emulated it were not simply powerless price-takers.[5] Their actions set an example for others to follow and helped people to survive.

The unwillingness of Africans to act in full accordance with European designs was also evident in the fact that several traditional handicraft textile industries endured the colonial encounter. Cottage industries were a source of significant demand for seed cotton throughout the colonial era, and were particularly strong in the cotton-producing zones of Afrique occidentale française. There, parallel or informal markets offered higher prices than the official market. This underground economy undercut the intentions of the French to maximize the volume of lint destined for Europe and to clothe their colonial subjects in re-imported cloth. Cotton that was bought and sold in these markets became a raw material for cultural maintenance and symbolic resistance. It was often crafted into traditional fashions and adornments that were used for aesthetic, ritual or performance purposes (Bassett 2001). While standardized European textiles were price competitive, West Africans considered these items to be second best.[6] Many amongst their number selected and purchased cloth not solely on the basis of its price, but also considered fabric quality, print characteristics and national origin. As textile shoppers are wont to do the world over, they also thought about the potential impacts that their choice to purchase and wear particular items could have on their social status. As people continued to prefer homespun goods the norm of Africans wearing African clothing held in the West.

Cotton colonialism nonetheless left a very European imprint on rural Africa. The top-down governance of cotton entailed the creation of a range of bureaucratic agencies and organizations to monitor and control the countryside. Approaches that these institutions adopted towards managing cotton did not change much after formal political independence (Bates 2005a). From Kenya's White Highlands through Tanganyika's killing fields and down to ground zero of the Mozambican debacle, forced resettlement schemes for export-oriented production trumped indigenous land tenure systems and land-use rights (Bates 2005b). Agriculturalists across the continent were forced, coerced, told or paid to adapt their daily lives to the exigencies of political programmes that aimed to promote economic efficiencies in the metropoles. Often the demands of these systems compromised the 'efficiency' of African life ways and sustenance strategies that had previously been viable. Many people starved as a result of the move from subsistence polyculture towards cash cropping that cotton was used to expedite. The apparent numerical success of production across the region during the late-colonial period – the rise of Sub-Saharan Africa's share of world cotton output from 1 per cent in 1927 to over 4 per cent in 1959 – must be understood in light of the external political domination, wealth extraction, underinvestment and hardship that were the gloomy corollaries of this growth. Europeans engineered and enforced a static export model that contrasted sharply with the dynamism that public actions and private power and innovation made possible down the value chain. Intra-European competition for supremacy in production and export markets did not lead to the transfer of spinning technologies to Africa (Baffes 2004a). Taken together, the outcome and after-effect of Europe's venture into Africa constitute some of the precursors to the inequitable aspects of the international political economy of cotton detailed below in Chapter 3.

However, the unfortunate and sometimes brutal experiences African farmers had with cotton did not always stem from externally-driven systems. North of the Sahara, inhabitants of rural Egypt were required to take up cotton under a homegrown plan prior to the colonial era. From 1805 through 1849, Muhammad Ali spearheaded a programme that made the cultivation of cotton with extra long staple lengths obligatory (Owen 1969). This campaign incorporated remote and previously self-sufficient areas into Ali's ambit. Through its total control of the market, the regime was able to capture a substantial margin between the prices that were paid to producers and the bids that were received from overseas cotton merchants. As such, the Egyptian government's autonomous introduction of cotton can be characterized as a simultaneous effort to achieve expansive dominance and extract rents. The principal difference between Ali's approach and later colonial systems was that he aimed to invest earnings from cotton in the formation and development of spinning and weaving industries on Egyptian soils. This move to ensure local ownership of the cotton textile industry antedated by over a century a wave of similar attempts to milk the African countryside and realize industrial development.

After formal political independence

After European colonizers ceded formal political control African governments attempted to gain autonomy from colonial trade patterns and industrialize their economies. Post-independence policies were shot through with the idea that markets needed to be governed to meet this dual objective. States created protective havens for their infant industries. Many erected tariff walls, removed duties on imports of industrial inputs, provided supply-side subsidies and directed credit to priority sectors. Where foreign investors were welcomed, governments also imposed local procurement and technology transfer requirements. States obtained grants from external donors and also took out loans from overseas creditors at market or concessional rates to expedite the development of energy production and distribution capacities, and for the upgrading of rail, road or port infrastructures. Exchange rates were also generally pegged to the US dollar at a high level in order to facilitate imports of intermediate and capital goods. Members of the political class rhetorically justified the latter policy choice as a necessary step on the path toward the realization of their aspirations for structural transformation.[7] Overall, it was widely assumed that this core mix of trade, industrial and macroeconomic policies would enable industrial development if domestic resources were simultaneously mobilized. Many governments consequently made efforts to maximize their earnings from agricultural production and natural resource extraction and trade. The hope was that these additional resources could be targeted and disbursed to raise the level of investment in new value-adding activities.

In countries that produced cotton and pursued the formation and growth of downstream sectors including spinning, textile manufacturing and clothing assembly, seed cotton prices were typically set at levels that were relatively lower than the prices paid to cotton farmers in other parts of the world. The region's low cash crop production costs – an offshoot of colonial conquest that supposedly constituted the region's static comparative advantage – were partly at the root of this harsh reality. Stingy producer prices were also a political choice (Bates 2005b). Governments generally chose to assist the efforts of their budding industrialists to keep down the cost of material inputs rather than maximize returns at the farm gate. They believed that the former preference would encourage investments in transformative development, an outcome that higher farm gate prices could not directly enable on their own. In so doing African states mimicked strategies that had been pursued elsewhere to protect cotton industrialists (Wolcott 2005). After colonialism then, sharecroppers, smallholders and tenants were subjected to what Colin Leys (1975) has justly labelled a 'political economy of neo-colonialism'.

Further to the point, many new governments left the colonial organization of seed cotton markets largely untouched during the immediate post-independence period. Markets for cotton retained their single-buyer character. Control was exerted over all aspects of production and ginning directly through crop boards, ministries or parastatal entities. As such, most African states did not have to resort to imposing direct taxes or quantitative restrictions on lint exports to ensure that their domestic spinners took delivery of product that cost well below the average world lint price. At least on paper, the day-to-day work of these organizations or their agents aimed to keep lint costs down. Where production levels exceeded domestic spinning capacities, or in the event of a sharp decline in the world price, low producer prices helped to ensure the price competitiveness of African lint on the world market. They also facilitated the efforts of boards and parastatals to cover the expenses of procuring and distributing productive inputs. Additionally, these prices made it easier for the monopsonies to deal with the costs of the relatively less efficient aspects of their operations, such as extension service provision, marketing, transportation and ginning. World market conditions also gave states a further rationale to maintain systems that effectively sanctioned the impoverishment of their cotton farmers. The amount of cotton available to be traded rose significantly over this period and global supply consistently outstripped demand. The extent and duration of market saturation was exacerbated by the fast growth of demand for synthetic fibres. The latter phenomenon reduced the share of cotton in the world fibre market considerably.

With one notable exception the preservation or copying of colonial practices held even in countries that did not produce significant amounts of cotton. In Ghana, the urban population's simmering demands for industrialization actually led the government to promote the development of a protected textile industry before domestic cotton production had been widely established (Poulton 1998). Ghanaian spinners had to procure lint from traders that were active in the sub-region while the Ghana Cotton Development Board (GCDB) constructed ginneries and brokered the idea of this cash crop to inhabitants of the northern region around Tamale. After local lint volumes rose, the GCDB set seed cotton prices at uniformly low levels to bolster Accra's textile industry. Kenya was perhaps the only country where new political elites actively resisted the tendency to reduce the prices paid for cotton. A highly concentrated post-independence redistribution of the productive lands formerly under European control ensured that the powerful and connected obtained significant tracts. This skewed process created a new land-owning class that had a vested interest in the realization of higher prices. According to Robert Bates, Kenya's new landed gentry articulated opinions on these prices that contrasted sharply with the standard view expressed on the 'need' to transfer incomes and wealth from 'backwards' agricultural activities to the cities.

In West African states the colonial legacy of state-controlled, single-buyer agricultural markets enabled governments to expedite rural-urban transfers. Parastatal successors to the Compagnie française pour le développement des fibres textiles (CFDT) such as the CMDT in Mali and the CIDT in Côte d'Ivoire organized the production, marketing and ginning of seed cotton. With total or majority ownership of the parastatals governments were able to appropriate any surpluses that these entities generated (Bates 2005a). Most cotton companies continued to set producer prices by fiat. Competition in the market for cotton only occurred in areas where unauthorized indigenous handicraft industries had survived protracted assaults. The unrelenting campaigns of the CFDT to eradicate unofficial trading had previously driven this economy underground or worse. In Côte d'Ivoire the CFDT's assertion of absolute control over the cotton economy entirely wiped out the activities of the Jula, once the most prominent cotton merchants in the sub-region (Bassett 2001). Marketing boards in the former British West Africa also assumed many of the characteristics of vertically integrated firms. For example, after 1968 the GCDB was responsible for all aspects of Ghana's seed cotton market. The board established seed, fertilizer and chemical distribution systems, and provided extension advice, ox ploughing and donkey carting

(Poulton 1998). However, the GCDB did not cover or subsidize the provision of these inputs and services consistently. The board typically charged its costs against the prices that growers were paid for their harvests.

Even so, parastatals and boards generally relied on government subsidies to assist their efforts to scale up production while they kept farm gate prices low. From 1970 the CIDT attempted to intensify production through the distribution of yield increasing inputs and the state reimbursed the company for the costs of the new package (Bassett 2001). Nearly all organizations that controlled cotton markets also obtained state funds to purchase and distribute subsidized fertilizer. The use of synthetic fertilizer across the West African sub-region was heavily concentrated in the cotton sector, and on average, states covered between 30 to 80 per cent of the import costs (Kherallah 2002). Companies and crop boards were consequently able to provide agriculturalists with an incentive to switch into cotton production even if the earning potential of cotton *vis-à-vis* other crops was unknown, ambiguous or only marginally better. Schemes to supply cut-rate fertilizers to cotton cultivators were by no means a net positive for producer livelihoods. They could enable beneficiaries to sell more kilos at the market, but could also be used for political ends to reinforce the notion that cotton farmers had a relatively good lot, induce inordinate reliance on cotton and rationalize low producer prices. John Baffes (2007) has argued correctly that fertilizer subsidies raised yields and were at the route of the vast production increases evident during the 1970s. The discussion below questions the distributional consequences of these systems and asks whether they enabled average cotton growers to reap significant gains. It seems that incentive problems along input distribution chains often skewed the allocation of subsidized goods and as a consequence, the distribution of benefits from the expansion of cotton output and sales in the west.

Governments and cotton boards in East Africa pursued similar strategies. Uganda's cotton board ensured that growers were paid well below the percentage of the world lint price that direct producers typically received.[8] This approach maximized foreign exchange earnings from lint sales and the government was able to draw upon these dollars to subsidize Kampala's industrialization. Earnings from cotton, tea and coffee exports were also the principal source of finance for the Nile headwaters hydroelectric project at Jinja (Bates 2005a). Likewise, the Government of Tanganyika invested hard currency earnings from its traditional export crop basket of cashews, cloves, coffee, cotton, sisal, tea and tobacco in the development of Dar es Salaam. The World Bank first articulated the need for an industrial push and the transformation of the agricultural sector through the 'establishment of highly capitalized settlements' in its 1961 *Plan for the Development of Tanganyika* (Hyden 2006: 120). The Bank had hoped that the new government would implement the study's recommendations, and the political elite subsequently lent its rhetorical support to these broad objectives (Chachage 1993).

However, the Bank's aspirations were never realized for more than brief periods. Prospective investors – both domestic and foreign – deemed the investment climate to be so unattractive that there was possibly a net outflow of capital from independence through 1967 (Bienefeld 1982: 299). That year President Julius Nyerere issued his Arusha Declaration. He embraced a socialist and autonomous strategy that ostensibly aimed to raise the quality of life in the countryside. Thereafter, official development assistance (ODA) flows and agricultural export earnings supplied most of the hard currency that could be drawn upon to fuel industrial development (Nyerere 1973). In this context Nyerere launched a parallel campaign for socialism and the restoration of pre-colonial family structures and rural life ways known in Kiswahili as *ujamaa*. Despite its ethical intentions *ujamaa* did not lead to increases in the production of traditional export crops (Pratt 2000). It eventually fostered a significant amount of forced internal migration and resettlement and also led to the imposition of strict controls over who was allowed to produce what, when and how. Development researchers later debated the extent to which this failure was caused primarily by a 'villagization' scheme implemented from 1973, or by factors such as climatic events, a lack of policy coordination or a dearth of resources.[9]

Beyond this debate, Tanzania's industrial development plans were impeded during the *ujamaa* period. Foreign exchange earnings from agriculture underperformed. While structural changes were evident and an uptick in economy-wide input-output linkages occurred, chronic hard currency shortfalls made Tanzania's already highly geared industrialization even more tenuous. In Manfred Bienefeld's view, even small declines in agricultural export earnings required the realization of substantial offsetting efficiencies in the industrial sector. Given the low level of human capital these enhancements were difficult to achieve. My interviewees highlighted several obstacles to investments in imported technologies or skills development abroad during this period. They argued that lending practices were often politicized and unmonitored, that covert consumer goods imports had fuelled foreign exchange leakage and that particularly unscrupulous elites had made covert deposits in dollar accounts they held at European banks.

Over this period Tanzanian cotton producers whose earnings placed them in the bottom two income quintiles fared relatively poorly. The most specialized amongst their number reaped only meagre gains – if any at all – from a small shift in the share of national income from the top quintile to the bottom 40 per cent that occurred between 1967 and 1972 (Green 1974: 268; Sandbrook 1982: 5). From the 1966–67 cotton-marketing season through 1972–73, cotton production volumes stagnated and trended downward and the proportion of lint export prices direct producers were paid declined significantly (Bienefeld 1982: 308). Regarding the latter trend, prior to 1966–67, growers had received payments that were equivalent to 65 per cent of the prices international traders or their agents paid the cotton board for lint on free on board (fob) terms.[10] This percentage steadily weakened down to the cotton price spike of 1973. In 1974–75 when world average lint prices returned to earth and local production dipped, farmers were only paid 41 per cent of the fob price (CRB 2004: 66; Bates 2005a: 138). Production volumes and administered farm gate prices remained relatively low to the end of the decade.

If John Saul (1974: 362) was correct when he contended that villagization was directed more at remote rural zones than at relatively 'advanced' areas such as Kilimanjaro, then it is also possible that cotton producers suffered disproportionately from physical and social dislocations and from food crop failures after 1973. The most remote farmers in Mwanza and Shinyanga regions certainly experienced hunger and starvation during the extreme drought of 1975. It is probable that their hardships were more intense than those that the powerful members of the politically connected Chagga ethnic group faced to the east. Domestic and international relief efforts were concentrated in the Arusha region, the homeland of the Chagga (United Nations 1978). Even to the southwest where villagization was more advanced food security was by no means assured. For example, despite the poor rains, a local bylaw in Kigoma obliged agriculturalists to plant a minimum acreage of cotton or face the prospect of imprisonment (Bryceson 1982: 564). Given the above trend, it is likely that average cotton farmers under *ujamaa* endured adversities that were markedly similar to or even worse than those that their contemporaries in West Africa suffered during the early neo-colonial years.

Across Africa cotton monopsonies experienced rising costs and other inefficiencies during the second decade of nominal local control that reduced their capacity to perform their functions equitably or with a semblance thereof. The year-on-year operating expenses of parastatals and boards increased significantly and were disproportionate to the noteworthy expansion of cotton production that occurred (Shepherd & Farolfi 1999: 7). The pace of these rises was also out of whack with the generally low level of demands governments and ministries made of these organizations to assume new marketing, procurement, service delivery or regulatory responsibilities. Critics of government intervention during this era have argued that politics were at the root of spiralling managerial and administrative costs (Leys 1996: 88). An interviewee alluded to this perspective when they asked if the evident growth in numbers of secretarial staff and chauffeurs at one board during its supposed zenith had been necessary for cotton production, a political production or the board's reproduction. Cotton boards and companies that were simultaneously regulators and market participants had incentives to engage in political efforts to increase their status or reputational capital relative to other state-owned enterprises or crop boards (Beddies et al. 2006). Whether to meet this end, or to justify their largesse, augment their capacities, advance the purposes of the organization or maximize the opportunities for managers to realize personal financial gains legitimately or otherwise, many took on tasks that overlapped with work that was being done by ministries of agriculture and research institutions. In so doing they generated coordination failures and resource misallocation. Amongst other outcomes, these flaws reduced the funds available to pay or provide services to the people who actually grew cotton.

Boards and parastatals also encountered problems achieving the timely delivery of chemical pesticides, a productive input that was then considered to be essential for the achievement of higher cotton yields. After transportation costs rose in 1973 prospective pesticide buyers in cotton-producing countries that did not export oil often experienced considerable difficulties securing enough hard currency to purchase adequate volumes from abroad at the outset of each planting season. They also had trouble paying upfront for the distribution of these chemicals and with maintaining and operating in-house transport fleets. These cost pressures had considerable ramifications. Coupled with the inadequate state of rural roads they impeded the delivery of pesticides to farmers. Accordingly, they were amongst the factors that precluded the timely distribution of pesticides and reduced seed cotton outputs in affected zones. Other aspects of the distribution problem included the makeshift and insecure nature of storage facilities along the chain and the fact that systems to ensure that employee incentives were aligned with the official and professed purposes of the organization were not generally in place. These realities enabled opportunists to appropriate and misallocate inputs.

Insiders that took these goods distributed them to their families and friends and several became illegitimate entrepreneurs. To move their hauls the latter created unauthorized secondary markets and made covert sales to farmers or dealers who were willing to pay (Bates 2005a).

Where parastatals and boards did not import and distribute pesticides directly their suppliers and distribution agents often secured contracts through patronage or other non-economic criteria. Businesses that were well connected, moneyed or had favours to call in were awarded contracts through tendering processes that were corruptible. Many of these firms were able to capture and retain their positions year-on-year irrespective of performance (Jones 1987). Distribution agents also often executed their contracts poorly due to a lack of oversight and accountability, shortcomings that were conducive to rent-seeking behaviour. Those involved with distributing inputs that did materialize at the village level often had inadequate capacities or incentives to ensure their equal allocation or to screen recipients to determine if they were indeed legitimate cotton growers.

While the latter problem was not evident in Tanzania, village-level input diversion did occur there. From the late 1960s male heads of local producer associations known as primary societies received inputs from a distribution system under the authority of the cotton board (Saul 1973; Gibbon 2001: 391). These volunteers were selected by village governments to distribute inputs directly to producers. They had strong knowledge of the volumes of cotton other male and female members of their society had produced and of the amount of inputs that these people required. Incentive problems nonetheless plagued the earmarking of inputs and deliveries. Primary society leaders were often the largest cotton growers. As the resources at their disposal increased the prospect that they would succumb to the temptation to divert to their own fields more chemical pesticides than they would otherwise be allocated became ever greater. Village officials and society leaders struck forward bargains on the quantities to be held back for their own use and also colluded to establish resale markets.[11]

Colin Poulton (1998) has shown that resources were also limited to monitor how the growers that actually received inputs subsequently made use of them. One consequence of this lack of supervision was that inputs that made it to the field were also redirected to other uses. For instance, many Ghanaian farmers registered for a unit of cotton with the GCDB simply to obtain a supply of fertilizer or pesticides that could be covertly applied to their subsistence or cash crop plots of white maize, or sold their inputs to other agriculturalists informally. Such clandestine sales were often made late at night. This diligence persisted despite the lack of oversight due to a pervasive fear of the authorities. Informal markets were sustained by equally strong anxieties over the prospect of food or cash crop failure. In years where it was thought that the cotton harvest would be poor or it was known that the cotton price would underperform the price of maize this exit option was especially attractive. Farmers sought out additional inputs in order to pre-empt prospective shortages or to cover at least some of their losses via the intensified husbandry of maize. Weather permitting, the latter crop guaranteed returns of both food and cash. The price elasticity of cotton acreage was accordingly high. Many that chose to defect in this way were able to re-register for cotton and receive inputs in subsequent years. Over time, those who established and maintained connections with players in official and unofficial input distribution systems secured increasing income streams and enjoyed greater levels of food security than those who were less well connected. The political economy of input systems thus fuelled a polarization of outcomes amongst producers in contexts as diverse as Ghana's arid northlands and the southern shores of Lake Victoria (Gibbon 2001). This was a truly pan-African condition.

Where credit subsidy schemes were operational these too encouraged a divergence in economic outcomes. In those places the disbursement of subsidized credit was then thought to be essential for the expansion of production volumes and the realization of higher productivity. This vision differed significantly from the approach advanced by micro-financiers in the present day. It aimed to make relatively sizeable sums available for the upgrading of productive capacities. Farmers that could access cheap credit facilities obtainable through agricultural development banks, credit associations or targeted government programmes were enabled to consolidate their landholdings and expand acreages under cotton cultivation. Others purchased additional inputs, implements such as seed drills, rakes and ploughs, and draft animal teams for their own use or for the hire or resale markets. Nevertheless, the gross value of the loans disbursed for these ends was often higher than the net amount available to recipients for productive investments owing to the fact that represent-atives and agents of the institutions that had provided credit typically took advantage of their intermediary roles. Many commanded prohib-itive fees to expedite flows of concessional finance. Evidence suggests that growers with the means to do so were able to bribe their creditors with relative impunity (Kherallah 2002). The few that benefited from this exclusionist approach to credit provision were also able to free ride on the lack of recourse lenders had to mechanisms that could ensure borrowers complied with the terms and conditions of their loans. Background information on personal credit histories was distributed asymmetrically, if it existed at all, and credit recovery efforts were scattershot and easily bought off. A culture of non-repayment and default was able to thrive in many remote cotton-producing areas where capitalist institutions and the means to enforce the rule of law remained underdeveloped.

Producers who did not enjoy reliable access to credit or other inputs received low prices at the market and also had to manage the reality that low prices were not the only pricing problem that limited their ability to reap gains from cotton. Prices offered for seed cotton were typically administered on a pan-seasonal basis and set prior to harvests. This form of market governance rendered prices invariable and restricted farmers' price-responsiveness. Under these systems there was rarely scope for buyers to offer growers incentives to deliver their seed cotton harvests to buying posts on predetermined marketing days. Had there been, companies could have been able to realize efficiencies and direct any savings that were generated towards raising average farm gate prices or to other organizational purposes. During harvest and marketing seasons poor growers consequently faced an unenviable choice. They could devote their time to the important task of tending food crops. Alternatively, they could pick, pack and schlep their harvest knowing that a low price awaited them regardless of when they chose to do so.

For those who subsisted on the margin of survival the choice between buttressing future food supplies and obtaining cash in the present was even less straightforward when cotton companies or boards were over-extended. At those times buyers or their agents could only offer impoverished producers promises to pay on a specified or ambiguous future date. In the extreme case of Tanzania it often took cooperative unions many months and sometimes several years to honour agreements to pay primary societies and their members. As companies and boards became more cash strapped over time, the possibility that farmers would not receive immediate payments became an increasingly bitter prospect. When there were no payouts to be had the trade-off between food and cotton was rendered into a raw deal. All cotton producers faced the unfairness of having to invest in and work for an ambiguous payoff. However, there was a clear differentiation between the degree to which well off farmers and their poorer neighbours suffered. The former experienced this condition as more of an inconvenience than an injustice. High-status households owned donkey carts or could hire them, and they could also employ day labourers to transport their harvests to market. These families coped relatively well when their buyers did not pay.

In West Africa pan-seasonal pricing and the inability of companies to ensure timely payments in a consistent manner also led to quality problems. There, the annual Harmattan windstorms that blew south from the Sahara from November through March posed a significant quality hazard. The longer that cotton remained unpicked during the Harmattan, the greater the possibility that the bolls would detach from plants and end up scattered all over the dry and dusty fields. This eventuality made the harvest more backbreaking and increased the probability that farmers would receive a discount for their 'dirty' cotton at the market. Even so, many farmers chose not to pick their cotton when immediate payments were not assured and prices remained invariable.

If yields were poor in a given season agriculturalists had a further disincentive to promptly harvest their crop. In addition to the factors recounted above, these unwelcome shortfalls also resulted from the failure of seeds, a lack of appropriate or even any extension advice, soil exhaustion and inadequate or excessive rainfall levels. Where they were encountered and uncertain payments dates were on the horizon women paid an especially high price. Each season across Africa women made unremunerated efforts to weed the crop and minimize the impact of this particularly intensive chore on their labour time. Consequently, the simultaneous occurrence of pricing issues and yield-reducing governance, market or environmental failures motivated many women to focus on their other considerable household responsibilities and on sustenance crops. In this context, intra-household obligations such as the procurement of water and wood, child rearing, health and elder care provision, animal husbandry, food preparation and compound maintenance could take precedence over attention to cotton. This response aimed to ensure subsistence, but left untouched the vicious circle of low quality and low prices that the neo-colonial cotton economy generated endogenously.

Nonetheless, cotton producers continued to deploy low-level everyday strategies of resistance from the time that they took delivery of inputs through to the day that they marketed their crop. Many of the tactics they made use of were similar in form to the approaches previously detailed, and as before, were not drawn upon uniformly across the diverse African countryside. Farmers did not enjoy equality of access to the means of opposition and the spoils from these minor struggles to capture higher returns were not distributed equally. Only one outbreak of collective resistance briefly emerged during the mid-1970s. At that time, several Tanzanian producers began an informal export crop production strike and threatened to withdraw entirely into locally-oriented cropping systems that they could control themselves. According to the late University of Dar es Salaam sociologist C.S.L. Chachage, these agriculturalists openly demonstrated their opposition to cash cropping and to their subservience to domestic and overseas markets. While this refusal was transitory, and only a fleeting, marginal factor in Tanzania's foreign exchange crisis, the collective organization that it entailed contrasted sharply with the individualist orientations exuded by large farmers elsewhere.

After independence no widespread revolt against the low prices and transfers of incomes and wealth associated with cotton production occurred. Input and marketing systems that engendered side-payments to influential producers gave these powerful figures an interest in the maintenance of informal schemes. The accessibility of rents could explain the evident dearth of attempts that were made during neo-colonial times to organize cotton producers to advance their collective interests (Bates 2005a). A new and powerful nationalism in many cotton-producing states that had not embraced multi-party democracy could also account for the reticence of average growers to challenge the rural order. So too could the fact that those who were directly responsible for cotton production and involved in a daily struggle to meet the rigorous demands of household sustenance – women – were typically disempowered within rural hierarchies and did not command the means to mount a durable challenge to the aspects of their exploitation that were associated with cash cropping. Patriarchal cultural and religious norms often ensured womens' obedience and subservience. Many frequently ceded control of their cotton crop and the meagre resources derived from its sales to their husbands. Acceptance of the status quo and compliance with the demands of official and traditional authorities can also be understood as a self-help strategy. Cotton farmers often adopted acquiescent postures in their dealings with authorities to ensure the security of their land tenure, assets or person, or to guarantee continued access to common resources, augment their social stature or build favour in the community. Whatever the particular causes of conformity were in particular contexts the overall result is clear enough. Empowered producers, input suppliers and functionaries at the companies, boards and ministries captured a disproportionate slice of the profits from the cotton economy.

Low prices and lopsided production, marketing and service-delivery systems persisted to the dawn of the 1980s. The political priority of industrialization also continued to squeeze cotton producers and their husbands, families and communities. Governments lacked the will or capacity to ensure that rural dwellers enjoyed the access that increasing numbers of city folk had to capability-maintaining and augmenting services such as clean water, education, electricity, healthcare and sanitation. States generally underinvested in the development of rural infrastructure (Havnevik 1993). National food crop surpluses were also subject to maldistribution and even to gross misuse (Bienefeld 1986; Leys 1996). These painful realities built upon the inexorable alteration of entire societies and life ways that commenced with the introduction of cotton monoculture and the money economy and was sustained through the implementation of agricultural techniques and technologies imported during the colonial era. Cotton producers were especially hard done-by. They continued to be advised to use methods that undermined the long run productivity of African soils and their own health more so than the approaches to the cultivation of other annual and perennial export crops that were adopted (Bassett 2001). Cotton growers depleted their natural capital more rapidly than their neighbours that had steered clear of cotton during this period. Taken together, these farmers faced the prospect of immediate and perpetual income poverty, food insecurity, social exclusion and capability deprivation. Relative to urban people, well-connected agriculturalists and even their husbands, they were considerably worse off.

Cotton sub-sector reforms and poverty in Tanzania

The Government of Tanzania was a reluctant and late adjuster. From 1979 when President Nyerere first approached the International Monetary Fund (IMF) to obtain a stand-by lending arrangement, Tanzania garnered pan-African repute for its strong stance in favour of sovereign policy autonomy and its opposition to the new conditionality. Nyerere was compelled to pursue this course only after it became clear that recourse to IMF funds was essentially the only way he could obtain the dollars necessary to correct a serious hard currency shortage. The oil price increase, unfavourable weather, an expensive war to remove Uganda's General Idi Amin from power, a costly nationalization drive, covert dollar-hoarding and a failure to rein in the appreciating exchange rate had induced a foreign exchange crisis (Helleiner 1999). An epic rhetorical battle ensued as Nyerere cut down the Fund's calls for demand restraint and refused to sign an agreement. Robert McNamara, then World Bank President, brokered the idea of a working group to find an accommodation over a year later. Both sides eventually rejected the output of these efforts to find a more balanced approach to stabilization and mitigate the impact of macroeconomic policy shifts on income distribution.

The Government subsequently put its engagement with the Fund on ice until Nyerere retired from the Presidency in 1985 (Biermann & Wagao 1986). A standby loan was finally agreed the next year and devaluation and the implementation of an expenditure-reducing structural adjustment programme rapidly followed (Harrison 2001). Measures such as the withdrawal of agricultural input subsidies, the introduction of user fees for government services and cutbacks in real allocations to the line ministries were pursued during the first phase of the plan (Kiondo 1993).

These reforms had a significant impact on the livelihoods of over half of the households in the Western Cotton Growing Area (WCGA) and the roughly 40 per cent of smallholder families nationally whose cash incomes were dependent upon cotton (Ratter 2005; Kabelwa & Kweka 2006). After the shilling was devalued and input subsidies were removed, prices of domestically produced and imported chemicals and implements rose consistently and significantly. Official seed cotton prices set by the Cotton Board – at that time the regulator, sole lint buyer and export monopoly – did not adequately mitigate these shocks. For example, the prices that the Board mandated the regional seed cotton buying and lint ginning cooperative unions to pay farmers during the 1986–87 and 1987–88 marketing seasons did not keep up with local price inflation. If any given member of a local cotton-producing primary society had wanted to obtain a sprayer pump, batteries and the standard package of agrochemicals they would have had to devote the shilling equivalent of 131 kilos of raw cotton to complete the transaction in 1986–87. The following season a similar purchase would have required the income earned from sales of 178 kilos (Meena 1991).

To put these figures in perspective, on average, women who cultivated cotton using hand hoes on the three acre plots characteristic of Mwanza Region were generally unable to realize yields that exceeded 150–200 kilos per acre (URT 1997: 47; The RATES Centre 2003). As the real value of the nominal prices these women or their husbands were paid for high quality (AR) or second-rate (BR) cotton fell, the typical levels of support that they were able to offer their six or seven family members came under considerable stress (Chachage 1993). Real producer returns were also diminished by the fact that the biggest buyers, the Nyanza and Shirecu cooperative unions, were typically unable to make cash payments at the point of purchase. It was not uncommon for the unions to delay payment of these IOUs for two seasons or more (Gibbon 1999: 131). Given that inflation was running at 30 per cent per year or higher this practice effectively short-changed many smallholders. At the same time it enabled the unions to take advantage of a substantial margin between the historical cost of the liabilities recorded on their balance sheets and the current value of their obligations to pay people (Gladwin 1991).

Inhabitants of the drought-prone Meatu District in Shinyanga Region were especially hard hit by these trends. The Sukuma agro-pastoralists had relied upon cotton as their sole cash crop for the previous two generations, and Meatu had become the most cotton-dependent district in the region. Cattle nonetheless continued to function as the principal store of value and source of social prestige. Prior to adjustment Meatu's cotton growers were subject to chronic food shortages and Sukuma children suffered from acute malnutrition. Despite the wealth evident in the 12 or 13 cattle that could be found on the large tracts held by each household, more people there than anywhere else in the region lived below the basic needs poverty line (TNDRDP 2004). At times of stress residents of this district were reluctant to part with their animals. If or when they attempted to do so they faced a lopsided buyers' market populated by only the very richest amongst their number. The cost of production and living increases that adjustment entailed reduced the viability of this already tenuous survival strategy. Poor people were forced to draw down their 'capital' *en masse*, a scenario that reduced returns from cow sales in the context of broader price rises. Declining real cotton prices exacerbated this vicious circle.

Peter Gibbon (1998, 1999, 2001), Colin Poulton et al. (2004a, 2004b, 2005) and their collaborators have exhaustively documented the processes that led to the Tanzanian experiment with cotton sub-sector liberalization and the outcomes of free competition. John Baffes (2002) has argued that the former can best be understood as a sequence of milestone events. The first market-oriented innovations came in 1991 when the Board enabled new firms to enter the market. That same year a new Cooperative Societies Act was signed into law. This act apparently aimed to sever the informal political links that had developed between the unions and the ruling Chama Cha Mapinduzi (CCM) party. It intended to stop cooperatives from transferring financial resources to various district and regional officials illicitly, and to reassert cooperative principles such as farmer control.[12]

A second landmark occurred during the 1991–92 cotton-marketing season when, as C.S.L. Chachage described, the Board set an 'indicative' price of 96 Tanzanian shillings (TZS) per kilo. In so doing it freed private traders that had entered the market the previous season to offer prices that they deemed fit. As it happened, many buyers disregarded the Board's indication and paid on average only 40 shillings per kilo. A few deployed avaricious strategies to procure cotton below the indicative rate. They made attempts to deceive sellers as to the existence of other buying opportunities and offered farmers immediate or more rapid payments if they accepted lower prices. The Government only intervened after the unions applied pressure and a minimum price of 56 shillings was established. Private traders were welcomed back the following season. That year the Board empowered the cooperatives and new buyers to set their own prices. Free competition was subsequently introduced during the 1993–94 marketing season. At that time the Board's export monopoly was delimited to the cooperatives' lint output and devolved to a new cotton marketing entity.

Private sector involvement nonetheless got off to a halting start. Deficient transportation infrastructure and a lack of credit opportunities constituted entry barriers early on. The Board had also resolved to deny new entrants access to cooperative ginneries. This orientation ensured that prospective buyers had to make costly investments in ginning capacity that they might not have originally planned to make. Firms that did enter also faced the vocal and openly hostile opposition of cooperative officials and employees in public and in the field. Even so, the simultaneous abolition of export licensing and foreign exchange surrender requirements in 1993–94 induced several investors to take the plunge. Coupled with exchange rate depreciation and the emergence of new lending opportunities from multilateral and bilateral creditors, these measures created enabling conditions for private investors over the ensuing years (Kanaan 2000).

Private ginners later gained toeholds as the regional cooperative unions came under increasing financial stress. Although the cooperatives continued to own over half of the rated ginning capacity down to the 1997–98 season, these organizations had become perennially dependent upon state-owned banks for the provision of ever-larger sums of credit. Their operations were shot through with patron-client obligations and corrupt practices. Chronic resource misallocation had led to readily observable deficiencies in ginnery and transport fleet maintenance and upgrading (Gibbon 2001). After the Board granted the unions export rights they were able to retain downstream agents to sell lint on cost, insurance and freight (cif) terms. The unions were consequently able to avoid the less favourable fob prices that their competition received and in all likelihood this reality kept these inefficient entities afloat for a time. It is probable that the relatively higher revenues per unit that the unions enjoyed were consumed by internal inefficiencies. Higher export prices were simply not passed along to direct producers. At the farm gate many were told to 'subiri kidigo sana' or 'wait just a bit' for the cash they were owed. As such the reputational capital of the cooperatives entered into protracted decline relative to some of the larger private sector upstarts that were able to offer cash at the market. The depreciation of the cooperatives' physical assets continued unabated.

Despite its advance billing as a panacea for the income shortfalls that had plagued seed cotton growers, the introduction and subsequent under-regulation of competition reduced the quality of the crop and compromised the global reputation of Tanzanian-origin lint. New traders were compelled to pursue aggressive strategies to pay back their loans and fulfil forward supply contracts. To do so, many focused solely on maximizing the volume of their seed cotton procurements (Gibbon & Ponte 2005). Buying operations increasingly ranged across distinct sub-varietal zones and no efforts were made to halt this practice or to effectively separate seed cotton varieties at the ginneries. This oversight led to the mixing of the national seed stock. Thereafter, yields and qualities were less predictable. Farmers were obliged to plant inconsistent seed mixes. These occasionally contained strains that were entirely untailored to local soil compositions or microclimates. As competition intensified the Board's capacity to monitor the quality of the harvests from mixed up seeds also broke down. Seed cotton quality was not graded effectively or even at all at the point of sale. These adverse outcomes reduced the scope for cotton farmers to reap the promise of liberalization.

The disappearance of grading reduced the incentive farmers had to keep their cotton clean and dry, and had a significant knock-on effect. Tanzanian lint had historically received higher prices on the world market for several reasons. Cotton there was picked entirely by hand, a practice that damaged fibres much less than mechanical harvesting. Roller gins were also predominant. These machines to separate seeds from the lint had a lighter touch than standard saw-type gins. Additionally, the annual timing of Tanzania's export readiness coincided with the low point of global lint availability. After quality grading was discontinued exporters increasingly faced the prospect that the premium associated with handpicked cotton would disappear. Quality discounts that they received for lint that was yellow-spotted or that contained high amounts of 'trash' or leaf matter could undercut or nullify the value of any premia they continued to receive for roller ginning or market timing. Tanzanian lint quality had declined so precipitously by 2001 that the International Textile Manufacturers Federation reported that this national origin was considered to be amongst the most contaminated in the world (Poulton et al. 2004b: 523). This reality reduced the capacity and willingness of exporters or their agents to offer higher prices at the farm gate. Given the secular decline of the world lint price and the emergence of new lint-producing and exporting countries, it also worked against the outlook that cotton could be made to work for the reduction of income poverty.

During the 1997–98 marketing season cotton farmer incomes came under severe stress. An input credit trust fund that had previously attempted to subsidize the pre-season credit needs of the lucky producers in the WCGA who had gained access to its resources collapsed. A surge of individual defaults and deficient credit recovery efforts were at the root of this especially ill-timed failure. Agrochemical prices had trebled over the previous four years, and most farmers had borne the full costs of insect-icides or had forgone the use of chemicals altogether (Baffes 2002, 2004b). Only two buyers offered credit or physical inputs on credit in 1997–98. Other firms claimed that they had insufficient cash on hand or too few resources to engage in the increasingly costly sourcing and distri-bution of chemicals. There was also a significant external disincentive for buyers to enter the input provision business (Gibbon 1998). Unscrup-ulous competitors could and did take advantage of the reality that cotton sellers were cash-poor at harvest time. Farmers needed little convincing to opportunistically defect from any pre-season obligations they had entered into, and sketchy buyers were all too willing to talk these people into parting with cotton that other buyers had a claim on. As one of my inter-viewees put it, these realities kept many buyers out of the business of 'taking responsibility for the input shortfalls of their suppliers'. While donors had disbursed concessional funds that had enabled the construc-tion of new ginneries, they did not extend similar support to remedy the input crisis (Gibbon 1999). They could not be expected to fund the new capacity *and* ensure its utilization.

A concurrent collapse of extension services exerted further downward pressure on the ability of average cotton farmers to raise their yields. Over these years fewer growers received annual visits from extension officers. Consequently, many did not have reliable access to new knowledge about productivity enhancing techniques. Few received a yearly reminder to avoid yield-reducing practices such as seed 'broadcasting' or the random scattering of seeds. Without this helpful advice many women did not take time out of other necessary household or food crop husbandry tasks to carefully sow or plant seeds in rows. They reverted to the easy, unproduc-tive approach.[13] In this context average yields fell as low as 500 kilos per hectare below the world average (Poulton et al. 2004b). At times when their seeds did not germinate or their plants succumbed to blight, farmers had to identify and remedy the problem on their own. Only those with sufficient resources or time could embark on a quest to seek the services of a district-level Cotton Board official (Kabelwa & Kweka 2006). Ironically, the breakdown of extension services flew in the face of an earlier multi-country World Bank study. This work had underscored the point that the realization of productivity improvements could improve crop performance and cash crop incomes more effectually than the reduction of agricultural taxation levels alone (Lele et al. 1989). Overall, the interruption of knowledge dissemination compounded the effect that price shocks had had on farmer attitudes towards cotton. As these factors reduced their capacities to make cotton pay, families began to prioritize other on and off-farm activities and looked for new income-generating opportunities.

Six years after liberalization it was difficult to argue that competition had had a positive impact on producer livelihoods. On the plus side, more farmers were being paid directly at the market for their harvests than under the regional cooperative monopsony system. Even though the volumes marketed by smallholders had declined, the greater availability of immediate cash payments fostered the impression that cotton production had become a more reliable and profitable venture. John Baffes (2002, 2004b) set out to ascertain the validity of the latter notion when he subsequently conducted an empirical comparison of pre- and post-liberalization farm gate and export prices. Baffes produced evidence indicating that growers had begun to capture a greater proportion of the prices that exporters received on the world market. The trend he identified seems to have held in the period immediately following the introduction of competition.[14]

Even so, the invisible hand alone was not at the root of this supposed 'win' for farmers. Higher prices were caused by the very visible withdrawal of the state and the provision of low-cost credit to inexperienced and inefficient entrants. As recounted above, the disappearance of input subsidies and extension services induced a quality crisis, yield shortfall and supply crunch. The shortage of quality seed cotton was not auspicious for new trader-ginneries. They did not have the equipment or know-how necessary to transform seed cotton into lint as efficiently as established ginners. As such these firms required more seed cotton volumes than the cooperatives to produce the same level of lint output. It was not the introduction of competition itself, but the inefficiencies of that competition and the woeful upstream productivity induced by liberalization that created the supposed sellers' market. It would take quite a lot of spin to argue that this outcome somehow vindicated the adjustment agenda or liberated farmers.

For the many growers whose production volumes and deliveries underperformed in the wake of liberalization, favourable market conditions were more apparent than real. Nominal price rises at the farm gate and spot payments did not necessarily raise the earnings of producers who marketed smaller harvests than before. Even for those whose nominal incomes from cotton increased, higher prices did not generally enable more consumption. George Kabelwa and Josaphat Kweka (2006) have argued that after an initial uptick, the basket of consumable and durable goods that could be bought locally with the shilling prices on offer did not expand and possibly even contracted. Moreover, the people who determined where or when to consume or what to buy were not often the individuals whose physical exertions, exposure to health risks and knowledge of crop husbandry had actually made the harvest possible. The persistence of a considerable gender imbalance in the consumption opportunities afforded by cotton ensured that within households, the benefits of nominal price rises were not broadly shared. The possibility of higher and immediate cash payments also incited many previously disengaged men to take charge of harvesting, head-loading and selling. This effect of nominally higher prices disempowered women who had delivered the crop to the market and controlled their family's income stream from cotton in the past.

Year-on-year the most well connected smallholders were also able to capture the most favourable intra-season prices. Where competition was intense, leading farmers who had dominated primary societies and who had been made better off by their control of these groupings (Gibbon 2001: 393) were differentially empowered. They were able to take advantage of the higher prices traders or their agents were willing to pay toward the end of each buying campaign. Those with the resources and the knowledge to hold back their own deliveries had an edge over smaller producers.[15] This outcome held even in places where primary societies had assumed a new role post-liberalization and acted as buying agents for the various local competitors. Under those circumstances, primary society insiders could encourage marginal farmers to part with their crop early and sell their own cotton late. They also secured and disbursed as they saw fit the fees the society earned for the organization and execution of weighing, record keeping, payouts and packing for onwards transport. In more marginal cotton-producing districts to the south and east of the Shinyanga and Mwanza regions such as Singida and Tabora, competition had a less discernible impact on prices. There, producers often faced single-buyer markets and at the outset of each planting season did not have certain knowledge that they would even have a buyer for their crop.

Despite these realities and other local particularities that compromised the ability of individuals to fight poverty, most analyses of reform have not explicated the totality of impoverishing relationships or articulated possible ways to alter or resolve distributional issues. In a 2004 article published in *World Development*, contributors to a six-country study of liberalized African cotton marketing systems only recounted and pre-scribed remedies for the sub-sector's lint quality, productivity and research ailments (Poulton et al. 2004b: 534). The authors carefully summarized the emergence of free competition and the consequent input market and quality crises, and also underscored the sharp 32 per cent decline in seed cotton outputs that occurred during the post-liberalization period down to 2002. Their account emphasized the evidently low-level of capacity the Cotton Board had to regulate competition, enforce the annual floor price or monitor and improve seed cotton and lint quality. They also high-lighted the Board's inability to resolve the input credit and provision fiascos[16] and effect a scaling up of research-related disbursements from the ministry or from donors. Colin Poulton and his team concluded that Tanzania's experience with liberalization had been an especially poor one and issued a call for buyers to lead renewal efforts. They suggested that the emergence of 'relational coordination' or informal private sector agreements might improve trust and be the way forward. In their view, the evidently low level of resources available to the Government meant that an endogenously generated private sector consensus on best prac-tices and the means to ensure sustainable seed cotton supplies was the only change-engendering mechanism possible. Under this proposal, the state would delimit its future work to the provision of public goods such as quality regulations and research.

Unfortunately, the market-based solution offered up by the team was built upon a questionable assumption. They seemingly presumed that all players in the sub-sector had accepted the ideal of liberalization and had an interest in the maintenance of competition. Brian Cooksey's (2003) conclusions on the illiberal nature of other Tanzanian export crop markets contradict such blind faith. Few private sector players have demonstrated thoroughgoing support for the so-called 'second-generation' reforms. They have shown little interest in the substantive transformation of corporate governance, and have not typically advo-cated greater adherence to international financial standards, pushed for safety nets or been wholly supportive of the national anti-corruption drive. As discussed in greater detail in Chapter 6, prominent private trader-ginners have also recently voiced anti-competitive sentiments. Suffice it to say at this juncture that conventional cotton buyers have expressed strong desires to socialize their costs to ensure a 'well-functioning' market, fill ginning capacities and boost margins. In a setting where the beneficiaries of 'liberalization' have actively opposed the liberal ideal and patronage still reigns, a prescription for more informal relationships seems wide of the mark.

The silence of the research team's otherwise comprehensive study on the broader context for the eradication of multidimensional poverty necessitates a fresh appraisal of how the factors that keep growers poor can be broken. Chapter 4 attempts to incorporate insights on the ways to break historic income-poverty maintaining relationships into a wider treatment of the measures necessary to overcome capability deprivation and social exclusion, and wipe out the lived experience of poverty.

The historic relationships today: A composite sketch

From the historicist account of the relationships between cotton production and poverty offered above it is clear that the wellbeing of most farmers has been under considerable and constant pressure. Many cotton growers have fared relatively poorly on each of the five aspects of wellbeing – material, physical, security, choice and social – defined through the World Bank's landmark participatory research initiative *Voices of the Poor* (Narayan et al. 2000). People turned to cotton out of physical or material necessity and did not often enjoy the freedom to choose an alternative crop or to exit cash cropping altogether. The cotton economy rested squarely on the labours of women whose access to income from seed cotton sales was frequently insecure. The personal security and status of these women within households and their social positions within communities have also been dependent upon their willingness to comply with the status quo. In areas where governance institutions and powerful males upheld the idea that cotton was necessary to pay taxes, generate an investable surplus, enable their own personal consumption or otherwise 'fight' poverty, women had few opportunities to dissent.

The impoverishing nature of the entire cotton enterprise is perhaps best demonstrated by the fact that farmers everywhere made attempts to exploit cotton systems in order to bolster their own wellbeing and enhance the livelihoods of their households and communities. While the affluence of a small number of producers fuelled the belief that cotton was 'white gold' and induced many others to continue to cultivate the crop, the notable successes of these individual farmers were the exception. Cotton fuelled increasing social differentiation that entrenched relative poverty as it compromised the traditional ways rural people had attempted to ensure their wellbeing. In Africa, cotton and poverty grew up together.

Today, a typical woman experiencing extreme cotton poverty cultivates and weeds her small plot of conventional cotton with hand tools and does not have access to labour-saving technologies such as ox-traction.[17] It is probable that she lacks formal property rights or title to the one or two hectares of cotton plants she tends to, or even to her subsistence food plots. Her opportunities to earn off-farm income are slim due to a lack of time or means of transportation. She is likely to be the principal food preparer, water and wood gatherer, and caregiver for any aged members of her household, for her young children or at times for her husband's other wives. She is also the probable go-to in the event that any of these people experience ill health. As she is preoccupied with multiple tasks, probably illiterate and lacking technical training her time horizons are short.

Consequently, she is not attuned to the future environmental impacts of conventional cultivation techniques or the concept of sustainability. She might be found to be drinking water from an old pesticide bottle while broadcasting seeds at the start of each season. If chemicals are available at the right time and she happens to procure scarce credit facilities to purchase them, in all likelihood she will spray her field with her feet exposed owing to the fact that she owns no shoes. Her access to organic or inorganic fertilizer is also slim to none, and she does not have the ability to ensure her entirely rain-fed crop against losses. At harvest time she gathers together a communal group to help with the cotton picking (a difficult task if she is a community 'outsider') and prepares a work-inducing feast for the team. With the crop in traditional baskets or plastic sacks, her husband takes charge of transporting the crop to a buying post. Having no means, such as a cellular phone, to compare the prices on offer at various locations, and in all probability having no access to anything approaching an effective producers' union, he is a price-taker at the point of purchase. The upside for him is that he controls the earnings from 'his' cotton and can proceed to purchase necessary goods for the household, or if he so chooses, spend a significant portion of the proceeds on an intoxicating local brew.

Theoretically then, a comprehensive analysis of cotton and poverty must cut across the four dominant academic approaches to defining and understanding the concept that Frances Stewart and her collaborators identified in a recent volume (2007). This empirical treatment necessitates attention to incomes and their distribution, to the relative and absolute deprivation of people's capabilities, to social exclusion at the individual and group levels and to the perspectives of farmers themselves. The next chapter hones in on the ways that the governance of the world trading system has impeded higher producer incomes, and on the limits of attempted reforms that have focused solely on the income dimension. Subsequently, elite and farmer perspectives on the multidimensional factors that must be overcome in order for Tanzanian cotton growers to reap a better deal from cotton are presented. Coupled with the above, these two discussions set up the ensuing evaluation of how globalization might be altering ideas, institutions and power relations and having impacts on poverty outcomes and the potential for poverty reduction.

3
Global Trade Governance and Cotton Dependence: Beyond Poverty Maintenance

Connections between international trade, cotton and poverty south of the Sahara are rooted in the economic structures bequeathed by the colonial era that persist to this day. Many countries in this region maintain a static specialization in cotton production and other diminishing return industries. This economic structure has been politically constructed and maintained over the past decades. Institutions and governance initiatives that governments of the South have launched to better their relative positions in the world trading system have been marginalized. This context has enabled a *de facto* incoherence between the trade and development policies of the European Union and the United States. The EU Common Agricultural Policy and US farm bills have maintained support systems for cotton farmers, ginners and traders that have led to increases in the global supply of lint and in the amount of cotton traded internationally. These policies have consequently exacerbated the general downward trend in the average world price offered for cotton lint and amplified the volatility of this price.

Several other trade-related factors have impeded the potential for cotton production to fight poverty and foster economic development. Many countries are more dependent on the export of lower value added cotton products today than they were during the early 1980s. Price rises associated with tight global supplies of maize, rice and other essential food commodities have also made subsistence more difficult for poor cotton growers that depend upon imports to meet their basic food needs. Many of these farmers also experience intense feelings of relative impoverishment. When their meagre incomes preclude the purchase of televisions or cellular telephones, or force them to rely upon low cost tradables such as used clothing or non-tradable foods such as cassava, poverty is exacerbated. The environmental costs of export-oriented production have also been impoverishing.[1]

Given these and other realities there are clear links between poverty, cotton production and world trade. This chapter focuses on the governance of trade and the entrenchment of the income dimension of poverty. It commences with an historicist account of how the international commodity trade has been governed since the end of the Second World War that links instances of malgovernance with poverty maintenance. After situating the progress of the Sectoral Initiative in Favour of Cotton at the WTO the chapter proceeds to analyse the limits of this commodity-specific proposal and its potential. In the concluding remarks a counterfactual point is then made on the eradication of poverty. It is simply unclear whether the opportunity costs of conventional cotton production and reliance on international trade are higher than those that would be associated with the scaling back or elimination of cotton cultivation and lint exports.

Global governance of the agricultural commodity trade and poverty

Governance of the global commodity trade was more notable for its absence than its presence during the era subsequent to the creation of the Bretton Woods Institutions and thereafter. In 1946 the United Nations convened a Conference on Trade and Employment to consider proposals backed by the United States and the United Kingdom to create an International Trade Organization (ITO). The ITO was slated to join the World Bank and the International Monetary Fund (IMF) as the third institutional pillar to govern the world economy. At the Conference a preparatory committee was struck to draft a Charter for the new UN specialized agency, and as the committee's work proceeded major trading nations began to negotiate tariff concessions in a separate process. Even though there were many debates in the broader dialogue on the ITO over the creation of International Commodity Agreements (ICAs) to stabilize and even raise revenues for countries that exported agricultural commodities, the tariff discussions did not focus on mechanisms for increasing world trade in agricultural products such as cotton.[2]

The following year the narrow tariff negotiations culminated in the signing of the General Agreement on Tariffs and Trade (GATT). As the GATT entered into force through a temporary protocol signed by 23 countries in January 1948, its backers assumed that it would constitute one component of the Havana Charter. This comprehensive document was signed two months later by 53 governments. However, the United States failed to subsequently ratify the Charter. Two years later Congressional enthusiasm had waned to such an extent that then President Truman had to concede that the ITO was a non-starter. As a result, a trade agreement that enshrined what today would be referred to as the policy space of governments to pursue economic development and structural transformation slid into history before any African colonies attained formal political independence.

Over the ensuing years a lack of governance continued to plague the agricultural commodity trade. In spite of some growth in its membership the GATT remained largely a club for rich countries. The priorities of the principal parties to the agreement dominated proceedings and were so at odds that the momentum of negotiations on tariff reductions was lost (Irwin 1995). This sluggish pace was assuredly not attributable to deadlock over agricultural trade liberalization. Discussion of the topic was virtually ruled out *a priori* as agriculture was not subject to regular trade disciplines (Coleman et al. 2004). GATT members did not have to apply principles of non-discrimination to the measures they imposed on agricultural imports. Governments were free from the national treatment obligation and could enforce stricter product standards on agricultural imports than those their national producers had to meet. As the most-favoured nation principle was also inapplicable, they did not have to apply any tariff reduction they might grant to a particular trade partner and member of the GATT to all parties to the agreement. The political choice to maintain the illiberal nature of the world agricultural trade evident in this regulatory freedom had the direct consequence of entrenching barriers to intra-North and South-North trade in like-agricultural products.

Confronted with a reality of double exclusion – both from the forum and from the substance of the discussions – governments that depended upon agricultural commodity exports were left with little choice but to make their case for redressing price and market access issues elsewhere. Several amongst their number raised questions about commodities at the anti-colonial Asian-African Conference held at Bandung, Indonesia in 1955. Later that year arguments were made at the UN General Assembly in favour of international mechanisms to stabilize commodity prices. Poor countries pushed hard for mechanisms to avert or mitigate export crop price shocks. At the Food and Agriculture Organization (FAO) in Rome consultations subsequently commenced on an international cocoa agreement (Gibbon & Ponte 2005). The adoption of General Assembly Resolution 1423 (XIV) several years later focused the attention of the UN Economic and Social Council (ECOSOC) on commodities. This resolution became increasingly relevant in the wake of decolonization as new commodity-dependent states joined the United Nations.

A North-South divergence over the prioritization of commodities was evident at the dawn of the 1960s. This divide was rooted in a clash of interests between Northern and Southern governments over the purpose of agricultural trade. The dominant view amongst Northern governments recounted by William D. Coleman was that agriculture was a weak, non-competitive sector that necessitated national management. In this light imports were only desirable if they increased national food security, provided consumables unavailable locally, or were necessary and low cost inputs for domestic or export industries. Southern countries that exported low value added products such as cocoa, coffee, cotton, jute, rubber, sisal and tea had an evidently conflicting interest. Their priority was to stabilize and maximize foreign exchange earnings from these goods. Many in the South argued that the objective of international efforts to administer commodity prices was to make prices more remunerative for producers. In their view, foreign exchange windfalls resulting from price increases could enable otherwise poor governments to obtain more of the intermediate and capital goods, knowledge and skills that they deemed necessary to expedite their pursuit of industrial development.[3]

Northern importers understood that efforts to stabilize commodity prices would benefit their buyers during times of constrained supply but rarely endorsed the establishment of global regulatory frameworks that aimed to stabilize prices. Where efforts to achieve price stability through commodity agreements were accepted, Northern traders were generally unsupportive of the developmental notion that prices should be set beyond market-stabilizing levels. They understood that such administration would lead to enduring increases in the costs of procuring the South's exports. As a consequence, industrialized governments typically rejected administered prices that would levy a development 'rent' with the potential to raise the costs of production amongst their established producers and thereby compromise domestic growth rates and employment. In effect, they sought ongoing protection for the rich and market justice for the rest.

Against this mercantilist stance the 'South' selectively deployed economic theory to make its case for governance innovation. Drawing upon the empirical work of Hans Singer (1950) and Raúl Prebisch (1950), political leaders argued that there was a secular or long-term downward trend in the terms of trade of economies that exported primary products relative to those that exported manufactured goods (Sneyd 2003; Toye & Toye 2003). Prebisch and Singer had noted that demand in the rich countries for commodities was inelastic and declining relative to the demand for machines and final consumption goods that were products of industrial processes (Sneyd 2006a). Politicians from poor countries also continued to flag political factors that constrained demand for their exports such as the high level of agricultural protection evident in the North (Singer & Ansari 1992: 75). Notwithstanding dismal academic evidence on the implications of commodity dependence that had come out towards the end of the 1950s, the South's activism on commodity governance ensured that it remained a prominent international trade issue into the next decades.[4] Over the 1960s and 1970s their push for commodity agreements and the redistribution of industry from the North to the South at times arguably became a matter of 'high' international politics (Sneyd 2005, 2006b).

The ideological conflict over commodity governance that took place after the birth of the UN Conference on Trade and Development (UNCTAD) in 1964 has been extensively documented elsewhere (Cox 1996a; Bhagwati & Ruggie 1984; Murphy 1984; Toye & Toye 2004). To gloss a few historic points relevant to the impoverishment of cotton producers, one decade after UNCTAD made the establishment of ICAs a high-level priority there was a palpable absence of new agreements. Southern governments continued to rely upon hortatory rhetoric to underscore the need for governance innovation. For example, in 1974 on 1 May the UN General Assembly adopted resolutions 3201 (S-VI) and 3202 (S-VI) on the declaration and programme of action for a New International Economic Order (NIEO). All but a few Northern countries were opposed to the measures necessary to realize the NIEO demands for a redistribution of wealth and wealth-generating activities to the South. Despite a simultaneous flurry of ad hoc declarations from African governments that asserted the need for renewed efforts to raise commodity prices, demands for higher prices were subsequently downplayed in the face of vociferous US opposition. The participation of the US in UN processes that aimed to address aspects of the NIEO package was secured thereafter. From 1976 the principal priority for commodity governance espoused by Non-Aligned Movement (NAM) members and articulated at meetings of the Group of 77 developing countries was the establishment of a special fund to stabilize commodity prices (Kaufmann 1989: 218).

The outcome document of UNCTAD's Fourth Ministerial Conference held at Nairobi in 1976 exemplified the readiness of the South to compromise after many years of stagnation. Resolution 93 (IV) on an Integrated Programme for Commodities did not include one objective that many individual members of the Group of 77 considered to be a key component of equitable commodity governance. The Resolution did not contain any language about their desire to engineer a system to index world commodity prices to the rising import costs that countries that relied primarily on commodity exports faced. Rather, Resolution 93 (IV) mandated negotiations on commodities under the integrated programme to work towards the more limited objective of achieving stable prices that were remunerative for producers and equitable for consumers (Toye & Toye 2004: 241). This vision for price stabilization was to be realized through the establishment of individual agreements for each of the commodities included in the programme. Buffer stocks of each commodity were to be allotted and managed such that they could be augmented to maintain targeted prices when supplies were abundant, and released on to the world market to curb prices when demand was strong. Resolution 93 (IV) also induced North-South dialogue on the creation of a Common Fund for Commodities to finance the start-up and operating costs of these price management schemes.[5]

Three years later negotiations for the fund were deadlocked and only one new ICA had been signed. Commodity negotiations subsequently lost steam as neo-classical economic thinking became increasingly influential. Global mechanisms to stabilize markets and facilitate redistribution were shunned as unduly interventionist and eschewed by the powerful new governments of the United Kingdom and the United States. Work on the integrated programme slowed considerably after these governments 'injected some realism' into the debate at the 1981 Cancún Summit on International Cooperation and Development. There, the US commenced a bilateral push to encourage the South to embrace 'free-enterprise capitalism' (Wiarda 1982). Due largely to deficient funding, existing commodity agreements came under pressure thereafter to abandon their price stabilization aspirations. In 1985 the International Tin Agreement collapsed and the cocoa and coffee agreements subsequently met similar fates. Only the rubber agreement survived for a time. These failures were partly rooted in the protracted international battle over the purposes of the Common Fund that delayed its establishment until 1989. They also stemmed from the objections of the US and several European governments to the principle of commodity price management or to the practical implications of ICAs.[6]

This latter problem in particular was at the root of the failure to establish an agreement for cotton. Six preparatory meetings of cotton-producing and consuming countries were convened during the late 1970s and early 1980s under the auspices of the integrated programme. At the first meeting the US representative voiced strong objections to the idea of any agreement that aimed to stabilize prices (Khan 1982: 289). From the point of view of the US government, fluctuations in the cotton price had been minimal relative to other agricultural raw materials and export earnings from cotton had been on the rise in poorer cotton-dependent countries. The US asserted that if the cotton price ever did warrant intervention, buffer stocks would be an inadequate price management mechanism.[7] It also considered an agreement for cotton to be infeasible due to the diverse qualities and grades of lint produced worldwide.

Most other participants opposed this stance. With the aid of technical assistance offered by the United Nations Development Programme (UNDP), 26 cotton-producing developing countries formed the Izmir Group. This group aimed to coordinate positions and counter US intransigence through enabling members to speak with one voice in the meetings. The Izmir Group strongly supported the creation of buffer stocks. It also pushed for the establishment of an organization to advance the provision contained in UNCTAD Resolution 93 (IV) to maintain and enhance the competitiveness of cotton *vis-à-vis* substitute man-made fibres such as polyester and rayon. According to Kabir-Ur-Rahman Khan's legal analysis, the US did not reject the idea that such a new institution was necessary at the outset. US negotiators actually argued that an institution dedicated to agricultural and industrial research was needed. They supported the first preparatory meeting's endorsement of a programme for research and development to strengthen the relative position of cotton on behalf of all producing and consuming countries.

In July 1977 the UNDP circulated a prospectus that elaborated detailed proposals for this programme, termed the Cotton Development International (CDI) (UNDP 1980). The proposed CDI aimed to absorb the small producers-only club, the International Institute for Cotton, and build upon that Institute's work to develop the world cotton market.[8] It also sought to expand the participation of developing countries in the marketing and distribution of cotton and to establish regional research and development centres. Attendees at an intergovernmental working group convened on the CDI at Geneva in October 1979 indicated broad support for the formation of a well-financed and comprehensive organization. However, the fourth preparatory meeting for a cotton agreement held in March 1980 had an inauspicious outcome. This setback occurred prior to the second gathering of the CDI working group. It ultimately delayed the drafting of articles of agreement and put preparations for a CDI founding conference on hold.

After the fifth preparatory meeting on cotton resulted in a further stalemate between the US and the Izmir Group, the Nordic countries tabled a minimalist agenda for the sixth preparatory. The Nordic proposal was for a sole focus on the issue of price stability. In the view of the International Cotton Advisory Committee (ICAC) Executive Director at the time, the Izmir Group was supportive of this objective (ICAC 1982). The US delegation did not demonstrate a similar level of accommodation. It refused to admit the need for price stability or supply control and the sixth preparatory was adjourned in failure. A paper on elements of a possible agreement drafted by UNCTAD in February 1982 failed to stimulate renewed dialogue. Its Secretariat subsequently spearheaded high-level consultations to salvage negotiations but these efforts gained little traction. Exasperated members of the Izmir Group later turned inwards to discuss their options. After 1983 the idea that a comprehensive agreement for cotton could be reached was increasingly seen to be an unrealizable utopian dream.

Despite this particular failure and the broader non-realization of the South's aspirations for commodity governance, the grievances articulated during the 1970s did leave an imprint on multilateral trade negotiations. During the Tokyo Round of the GATT it became clear that rich parties to the agreement did not entirely reject the principle that the South's exports should be subject to special and differential treatment (Winham 1984). At the conclusion of the Round in 1979 agreement was reached on a measure that allowed rich members to give poor countries preferential access to their markets. This so-called enabling clause sanctioned systems of trade preferences that did not require reciprocal concessions from Southern beneficiaries. Such a Generalized System of Preferences (GSP) had been discussed at UNCTAD since its formation. The *de jure* endorsement of the principle of non-reciprocal, discriminatory treatment in favour of poor country exports was a significant paper concession.

In practice, however, rich countries continued to exclude agriculture from regular trade disciplines and preferences were mainly applied to industrial products. The main beneficiaries of the GSP 'tended to be Asian countries with strong manufacturing' bases, owing to the parallel exclusion of textiles, clothing and other light manufactures from preference offers (Gibbon & Ponte 2005: 50). This system consequently underperformed expectations and reflected a narrow definition of special and differential treatment (Ozden & Reinhardt 2003). It remained far from the ideal 'double standard' for development that Swedish economist Gunnar Myrdal had previously deemed necessary for the South to overcome 'rigged rules' and pursue structural transformation (Myrdal 1970: 294). For the poorest non-oil exporting countries that were struggling to manage steep increases in their import costs at a time when the volumes and values of their exports were stagnant or in decline, the GSP was effectively irrelevant.[9]

As commodity governance initiatives languished within the UN system many developing countries moved to participate aggressively in the Uruguay Round of the GATT negotiations. From September 1986 their new engagement was fuelled by the inclusion of agriculture in the negotiations and the belief that agricultural liberalization in the North would occur after the successful conclusion of the Round (Wolfe 1998). The final agreement reached at Marrakech in 1994 had many developing country signatories. By signing the GATT 1994 and creating the World Trade Organization (WTO), parties agreed to be bound by all of the new agreements that had been negotiated. These included the agreements on agriculture, intellectual property rights, investment, sanitary and phytosanitary measures and services (Coleman 2005). Developing countries acceded to this single undertaking and to empowered mechanisms for the settlement of trade disputes. They perceived that the latter would be a useful means to attain better access for their agricultural products in Northern markets if developed countries failed to adhere to the schedules for liberalization set out in the Agreement on Agriculture (Gibbon & Ponte 2005: 54). According to Joseph Stiglitz these countries knowingly ceded policy autonomy under the WTO agreements on intellectual property and services in a 'grand bargain' to realize their decades-old market access objective (2006a: 77).

The Agreement on Agriculture mandated reductions in the ways that Northern governments subsidized their agricultural exporters. It also sought to significantly lower the aggregate levels of trade-distorting domestic support that rich governments provided to their agricultural producers. Additionally, import licensing and other non-tariff barriers (NTBs), quotas and restrictive health and safety standards were to be phased out. Despite the agreed objectives government support for agriculture in the North subsequently increased. The process of converting WTO illegal subsidies into tariffs known as tariffication proceeded haphazardly and in what Peter Gibbon and Stefano Ponte (2005: 55) refer to as an especially 'dirty' manner. Often new bound tariff levels – the highest possible rate that can be levied – afforded more effective protection to Northern producers than the subsidies or quotas previously in place. These defections ensured that Sub-Saharan Africa's share of world agricultural exports remained stagnant at about 2.5 per cent through the turn of the millennium, down from the roughly 8 per cent they accounted for at the outset of the Tokyo Round (USDA 2002: 2).[10]

In sum, the governance of agricultural commodities or the lack thereof has had a directly impoverishing effect on commodity-dependent countries. During the height of activism for the managed stabilization of commodity prices at more remunerative levels from 1950 to 1984, world agricultural prices declined in real, inflation-adjusted terms by over 1 per cent per year (Leys 1996: 140). These price declines were induced by a consistent oversupply that resulted from increases in agricultural productivity and exports. To cut their input costs and maintain their competitiveness, market shares or shareholder values, downstream industries consistently searched for and demanded cheaper supplies, and this too was a factor that reduced agricultural prices. As these commodities constituted a diminishing proportion of world trade, persistent specialization south of the Sahara entailed a drop in its share of world exports from 5.7 per cent in 1963 to 2.4 per cent in 2000.[11] This occurred despite the coming into force of the GSP, the Lomé Convention and other preferential schemes (Gibbon & Ponte 2005: 37). The declining share of African exports in rich country imports over those years also reduced the leverage these states had to effect governance changes.

The impoverishing realities of lower prices and diminishing influence were compounded by the fact that the real prices of Africa's principal exports were in long-term decline relative to their real import costs (FAO 2004a).[12] Since 1980, export revenues from agricultural raw materials have deteriorated and lowered the capacity for countries that rely on these exports to pay for their imports. Unfortunately, this was the exact outcome that advocates for the redistributive management of commodity prices had been trying to prevent (Birdsall 2007: 230). Malgovernance that enabled supply gluts and sharp reductions in export prices also spawned numerous economic contractions. Nancy Birdsall (2007: 237) has argued that these phenomena entail the disproportionate suffering of the poor. To compensate for governance failures, African least developed countries (LDCs) have pursued greater levels of access for their products in the North. In so doing these governments have done more than other developing countries to dismantle trade barriers (UNCTAD 2004a: 16; Kiely 2007). In 2001, high-income countries applied tariffs of 48 per cent on average whereas in low-income countries, tariffs averaged 17 per cent (Blouin & Weston 2005: 1). As detailed below, this relative openness reduces the policy tools necessary to pursue value addition and reverses the historical sequence whereby growth has typically predated liberalization (Rodrik 2007; Chang 2008). In effect, many of these countries have opted to govern their own trade policies in a manner that advances the interests of the North, just as the latter's interests have been assured through the absence of global governance for commodities.

Governing cotton at the WTO?

After the turn of the new millennium it was apparent that the Uruguay Round 'grand bargain' had become a raw deal for African cotton exporters. In 2001, the real world price for cotton lint, having fallen 66 per cent from 1995 levels, dropped to the lowest point recorded since the 1930s (Watkins with Sul 2002: 2; Weston 2007: 1). This disastrous outcome occurred despite the fact that the overall support that governments of cotton-producing countries provided their cotton sectors had declined. In spite of the worldwide trend, a number of countries maintained comprehensive cotton subsidy programmes. These systems raised export volumes and exerted significant downward pressure on the world price (Gillson et al. 2004). They remained in place even though WTO members had previously agreed to reduce agricultural subsidies that fell within the so-called 'amber' box of trade-distorting measures under the Agreement on Agriculture.[13]

As regards the dominant cotton exporter, the US effectively defected from these commitments and the further pledge to 'decouple' its subsidies from production after its 1996 farm bill was signed into law. On the latter point, the Agreement on Agriculture deemed price support measures in the form of administered prices or direct payments to producers to be trade distorting. The general understanding was that the presence, availability and level of these subsidies could induce non-price competitive farmers to grow crops that farmers in other countries could produce at a lower cost without state support. In the language of the WTO these subsidies were 'coupled' with the decisions of farmers to plant and were highly correlated with subsequent production volume increases. As such, they were an invitation to trade distortion if it caused domestic spinners to shun imports or if domestic demand for the subsidized product was insufficient and it had to be exported. If these eventualities played out, more efficient producers elsewhere could lose market share, receive lower prices or even be priced out of the market. The Agreement consequently mandated a decoupling of subsidies from farm production decisions. Policy changes contained in the 1996 farm bill did not live up to this objective. Cotton subsidies were only superficially decoupled. Rather than subsidize farm gate prices the US introduced a new scheme of 'production flexibility contract' payments for its cotton farmers. This system based the amount of support offered on historic yield and acreage levels. By re-branding price support measures in this way the US was able to foster the notion in its notifications to the WTO that it was complying with its commitments. Over the following years Congress also approved several large ad hoc appropriations for cotton producers. This additional support took the form of direct payments and belied the idea that the US had embraced the spirit of Uruguay. It induced cotton production and lint export increases that lowered the world price further still (Goodwin & Mishra 2006).

Subsidies in the EU remained similarly in force post-Uruguay. These measures accounted for roughly 16 per cent of the total world value of cotton subsidies and were trade distorting. At least one major study suggested that the $700 million spent by the EU annually on these production-stimulating measures did disproportionate harm to several West African countries that were dependent upon cotton exports (ODI 2004: 1; Baden 2004: 1). In 2001, the total cost of lint produced in Greece and Spain was three times higher than what would have been spent that year if production had been shut down and an equivalent amount of lint had been imported from Africa (Goreux 2004: 5). Countries such as Brazil, China, Egypt and Turkey also continued to subsidize their cotton producers to lesser degrees. Many LDC exporters also maintained WTO-legal input support schemes. However, the relative impact of the US and EU programmes on the world price was much greater (Stiglitz 2006a: 84). For example, the *World Development Report 2008* took on board the estimate that production and exports prompted by US subsidies of $3–4 billion annually have consistently reduced the world cotton price by 10 to 15 per cent (World Bank 2007: 99).

Exceptionally low world prices in 2001 led to losses in excess of 1 per cent of GDP annually in Bénin, Burkina Faso and Mali. These countries depended upon cotton for 5 to 8 per cent of their GDP and 30 to 40 per cent of their foreign exchange earnings (Watkins with Sul 2002). As Kevin Watkins noted at the time, Burkina Faso's GDP per capita was $237 in 2001, while each acre of cotton under production in the United States received $230 in subsidies that year.[14] Ironically, a few US producers had started to refer to their crop as 'poverty weed' during this period due to its high production costs and the low prices they were paid at the market (Kripke 2005: 75).

The 2002 US Farm Security and Rural Investment Act did not phase out the ostensibly 'decoupled' payments. Instead, it encouraged farmers to update the acreage and yields data that were used to calculate payments and re-labelled these 'direct payments'. The Act also offered farmers counter-cyclical payments that gave them a perverse incentive to produce more when world prices were low. Additionally, it renewed marketing loan and market loss assistance programmes. These measures respectively ensured that growers did not experience credit shortfalls and that they would receive a guaranteed price if they could not find a buyer (Goodwin & Mishra 2006). Payments offered to domestic spinners and exporters known as 'Step 2' were also maintained. This programme compensated lint buyers for the savings foregone when they purchased US cotton that was relatively more expensive than the prices they would have paid on the world market. Importers of US cotton were also extended export credit guarantee facilities. With this measure, the Act sought to enable US banks to offer more loans to foreign cotton buyers at rates that were much lower than they would have been able to obtain at home or in the US without the governmental guarantee (Sumner 2007: 5).

Overall, these programmes generated patently irrational and unequal outcomes. Regarding the former, investments of taxpayer funds in the sector have typically been greater than the private returns participants in the subsidized industry have earned from their output. Governmental support to US cotton producers totalled $3.508 billion in 2002, while the value of the crop marketed was only $3.497 billion (Sumner 2007: 21). The costs of US cotton production that year were fully socialized while the benefits remained privatized. This intensified the trend evident from 1998 through 2002 whereby the total cost of annual US cotton support was more than the total prices that were paid for US cotton exports (Goreux 2004: 22). Unequal outcomes associated with the programme were generated internally and also through its externalities. Within the system, disbursements were highly concentrated amongst the biggest farmers, ginners, spinners and traders.[15] Beyond the domestic disparity, the programmes undermined not only the livelihoods of some of the most income-poor people on the planet, but also the viability and inter-ests of lint export industries in richer and poorer countries alike (Stiglitz 2006a: 86, 307).

Simmering discontent amongst its exporters subsequently gave the new government of Brazil impetus to question US subsidies at the WTO. On 27 September 2002 Brazil requested consultations on the legality of US cotton support. In their analysis, the US had consistently increased the amount and value of the trade-distorting measures it deployed. As a consequence, Brazil argued that the US no longer enjoyed immunity under Article 13 of the Agreement on Agriculture. Also known as the Peace Clause, this article limited legal challenges to only those subsidy schemes that could be shown to have increased in depth or breadth since 1992. The Brazilians also considered the US system to be injurious and in contravention of Article 6.3 of the Agreement on Subsidies and Counter-vailing Measures. In particular, they argued that Step 2 payments to domestic users and traders had increased US production and exports to such an extent that the US had captured a greater share of the world lint trade and depressed the world price (Gillson et al. 2004; Benicchio 2005: 34). For its part, the US countered that it had maintained its support at 1992 levels and decoupled its payments from production. Brazil formally requested the establishment of a dispute settlement panel on 6 February 2003. On 18 March, the WTO General Council meeting as the Dispute Settlement Body (DSB) moved on their request.

Cotton-dependent African governments viewed these proceedings with interest from the sidelines. Bénin and Chad paid especially strong attention and subsequently took steps to become third parties to the dispute (Zunckel 2005). However, these two governments and other members of the African Group at the WTO did not consider the develop-ing dispute to be a silver bullet for progress on agricultural trade liber-alization. It was viewed to be a necessary supplementary approach to trade negotiations on the topic. The latter had been prioritized as part of the Doha Development Agenda articulated at the Fourth WTO Ministerial in November 2001. Nonetheless, progress in these nego-tiations had been halting at best. A Special Session of the agricultural negotiations held on 25–31 March 2003 might have been a tipping point. African representatives concluded from the evident impasse that modalities for further commitments to liberalize agriculture would not lead to the realization of special and differential treatment in any sub-stantive form. This awareness and the perception that negotiations had stalled led members of the African Group to entertain new ideas that aimed to transcend the status quo.

It was in this context that the idea of invoking special product status for cotton shot to prominence (Goreux 2004: 10). African discontent cul-minated on 30 April when Bénin, Burkina Faso, Chad and Mali submitted their request to treat cotton as a special product necessitating specific treatment to then WTO Director-General Supatchai Panitchpakdi (WTO 2003a).[16] In this document, entitled *Poverty Reduction: Sectoral Initiative in Favour of Cotton*, the four signatories called for a mechanism to phase out support for cotton. They also requested short-term compensation for all cotton-producing and exporting LDCs while the US and EU subsidies remained in place. The four backers of the Initiative, or as they later came to be known, the Cotton Four (C4) countries, specified the criteria for compensation and modalities for evaluating the levels of compensation required (Amehou 2005: 25).

Cotton became a key component of the Doha Round after the C4's intervention. Its place was secured despite the opposition of several trade economists to this course of action. The Initiative's critics considered it to be a distraction to progress in the wider agricultural negotiations and also played up the fact that the text of the Doha Declaration had not referred to the commodity specifically. The Sectoral Initiative was nonetheless enshrined amongst the principal African objectives for the Round. A dramatic and unusual presentation solidified this status. The President of Burkina Faso, Blaise Compaoré, presented the Initiative to the WTO Trade Negotiations Committee on 10 June (Zunckel 2005: 1079; Goreux 2004: 4). President Compaoré requested the design of a specific mechanism to 'progressively reduce' with a view to 'fully suppressing all cotton subsidies' (WTO 2003b). He advocated 'immediate and transitory' compensation for foreign exchange and revenue losses incurred due to the subsidies. Compaoré noted that the signatories were not asking for charity, preferential treatment or aid. In his view, the Sectoral Initiative simply sought 'the application of the free market rule'.[17] He urged attendees to take up the Initiative at the Fifth WTO Ministerial at Cancún the following September.

Several controversial discussions on the topic occurred during the lead-up to Cancún. The Initiative was raised during two separate meetings of the Agriculture Committee that July. From the discussions that took place it was clear that 13 other West and Central African governments supported the expedited treatment of cotton. They desired an 'early harvest' of liberalization commitments in this area. Several members opposed this stance and questioned whether cotton was indeed a special product. They raised the idea that it might be more fruitful to re-embed cotton under the three discussion pillars of the broader agricultural negotiations: domestic support, export competition and market access (Baden 2004: 4). The Initiative's opponents also expressed concern over the precedent that any compensation scheme would set, and inquired about the sources, viability and general need for such funds.[18] The C4 and its supporters strongly disagreed with these positions and worked to make cotton a principal focus of the Fifth Ministerial.

At Cancún, a pre-Ministerial event sponsored by Germany focused attention on the topic, and a Special Session of the opening plenary was devoted to cotton. Even so, specific references to the cotton proposals were not included in the negotiating texts. The controversial 24 August draft ministerial declaration and the revision of this document issued in the midst of negotiations on 13 September did not mention the Initiative's components. According to Sally Baden, strong opinions voiced by the US and EU delegations worked against the inclusion of these specifics in the revised draft. The EU and US rejected compensation, and characterized such measures as external to the existing prerogatives of the WTO (Baden 2005: 97).[19] In the hallways these delegations implored the C4 to diversify away from cotton. This imperative was highlighted in paragraph 27 of the revised draft text prepared by Luis Ernesto Derbez, then Mexican Foreign Affairs Minister and Chairperson of the Ministerial. Minister Derbez had attempted to strike a balance between a push led by the US and EU to move the Singapore issues (competition policy, government procurement, investment and trade facilitation) up the negotiating agenda and the aspirations of the newly formed Group of 20 (G20) for progress on agriculture. The African, Caribbean and Pacific (ACP) countries, the LDCs and the African Group that together comprised the Group of 90 (G90) backed the principal objective of the G20.[20]

After a consultative process that lacked transparency Derbez articulated his solution to the impending North-South conflict. He set out to create working groups on the topics of agriculture, development, non-agricultural market access (NAMA), the Singapore issues and other relevant matters. After this system was adopted it quickly resulted in the total breakdown of negotiations and led members of the G90 to quit the talks (Pesche & Nubukpo 2005: 51). During the weeks following the walkout, Bénin's Ambassador to the WTO, Samuel Amehou, worked in Geneva to mobilize other members of the African Group on the cotton issue. On 7 October Bénin submitted a redraft of paragraph 27 that argued for the re-inclusion of the Initiative's specifics (WTO 2003c). As a result of this activism, cotton, along with agriculture, NAMA and the Singapore issues became one of four principal foci of efforts to renew negotiations after the failure at Cancún.

While the new priority status of cotton stimulated much discussion on the topic within and beyond the WTO this flurry of activity effectively marginalized the C4's push for a transitional compensatory mechanism. On 12 February 2004 the European Commission announced a new EU-Africa partnership on cotton. The text of this agreement endorsed the WTO as the primary venue for negotiations on the reduction of trade-distorting support. It also promised that policy changes would be forthcoming to decouple EU cotton subsidies from production. Instead of backing the principle of specific compensation, however, the text floated the idea that a financing mechanism under the EU-ACP Cotonou

Agreement (the successor to the Lomé Conventions) could be revived and oriented to assist cotton-dependent countries facing balance of payments crises (Baden 2004: 9). The partnership also prioritized the scaling up of technical assistance to these countries to improve their cotton systems. Similarly, a joint OECD-DAC/WTO meeting in Paris on 2–3 March and consultations on 5 March in Geneva supported the idea that the opportunities for financial and technical assistance needed to be elaborated and expanded (WTO 2004a). The compensation issue was also not a focus of these discussions.

Later that month this topic was marginalized further still. An African Regional Workshop on Cotton organized by the WTO Secretariat and held at Cotonou, Bénin instead honed in on development assistance. Attendees at this gathering included the C4 and 26 other cotton-producing and trading African countries, 18 intergovernmental and multilateral organizations, and the US, EU and China.[21] Much of the dialogue at the Cotonou workshop concentrated on the division of labour necessary to mobilize effective technical assistance and on the mechanisms that could enable concessional financial flows for cotton to be increased. This event furthered the distinction between the trade and development aspects of the cotton problem (Amehou 2005: 26). In effect, it enabled the WTO to engage with members of the wider international development community without substantively addressing the issues of compensation or subsidies (Goreux 2005: 87). The event also afforded the EU and the US an opportunity to engage in public relations. For example, the US attempted to link its bilateral and highly policy conditional funding window, the Millennium Challenge Account, to the reduction of cotton poverty.

The post-Cancún Doha Agenda work programme agreed by the General Council on 1 August of that year entrenched the separation of the trade and development aspects. The text of the so-called July Package reaffirmed the importance of the Sectoral Initiative and outlined the parameters through which the trade aspect would be taken up within the agriculture negotiations. Cotton was to be addressed ambitiously, expeditiously and specifically with respect to all three pillars of the negotiations (WTO 2004b). A sub-committee on cotton was also called for to meet periodically and report to the Special Session of the Agriculture Committee. This body aimed to ensure that the General Council's wish to prioritize the issue was fulfilled. The July Package also 'attached importance' to the development aspect. It mandated the Director-General to take the lead on the mobilization of resources, implored members to focus on this topic and instructed the Secretariat to continue its engagement with the development community to realize greater levels of assistance. Paragraph 5 of the annex asserted that coherence between the trade and development aspects of the cotton issue would be pursued solely through the means detailed in the package. As the C4 proposal for specific compensation articulated in the original Cotton Initiative was not referred to in the July text, this paragraph effectively ejected the second core feature of the Initiative from the Doha work programme.

While initial actions were being taken to implement the July Package, the panel report on the Brazil-US Upland Cotton dispute was released on 8 September. The panel ruled largely in favour of Brazil. It found that price-contingent payments offered by the US, including counter-cyclical payments, marketing loans, market loss assistance, and Step 2 payments had significantly depressed the world lint price. In the view of the panel these measures constituted serious prejudice against Brazil (WTO 2004c; Benicchio 2005: 35).[22] It also ruled that Step 2 payments to domestic users contravened the Agreement on Subsidies and Countervailing Measures. These payments to US-based cotton exporters were considered to be illegal export subsidies principally due to the fact that they were contingent on the purchase of a product containing domestic content. The export credit guarantee programme was similarly ruled to be a prohibited export subsidy scheme. The panel also targeted production flexibility payments and direct payments that the US had categorized as minimally trade-distorting green box measures in its notifications to the WTO. It found that this type of support precluded the cultivation of other types of crops on land eligible for subsidies. As such, the panel ruled that the US had breached the Peace Clause and that these measures were amber box subsidies subject to the reductions mandated in the Agreement on Agriculture.

With trade negotiations ongoing, the US subsequently appealed many of the panel's findings. The Appellate Body ruled on these appeals on 3 March 2005. It dismissed all of the US claims and largely upheld the panel report (WTO 2005a). Meeting as the Dispute Settlement Body (DSB), the WTO General Council adopted the panel and Appellate Body reports on 21 March. The DSB instructed the US to respond to these rulings but did not stipulate a deadline for the removal of the offending programmes.[23]

One month after the reports were adopted the African Group sought to build upon the evident momentum for the removal of offending support. Then European Trade Commissioner Peter Mandelson had fostered the impression of progress when he recommitted to decoupling European subsidies during a 13 April speech. The African Group subsequently proposed new working methods for the negotiations and mechanisms to realize progress on both aspects of the cotton problem. On the trade aspect, it called for the North to open its markets completely to African cotton and derivative products. The Group advocated the removal of export subsidies by mid-year and distortionary domestic support by 21 September (Amehou 2005: 28). Regarding development, the African Group prescribed the approval, design and financing of an emergency support fund for cotton. It wanted this special fund to be in place by the end of the year and lobbied for the fund's payouts to be set at a level equivalent to the value of 20 per cent of the cotton produced by African countries (Baden 2005: 97). This push reasserted the imperative of compensation in a new guise, but largely fell upon deaf ears. On 18 July, as the world price hit a new slump, Bénin condemned the lack of progress on all fronts of the African Group proposal during a meeting of the cotton sub-committee. Despite growing political and public awareness of the issue, little forward movement was made on the African objectives in the lead up to the Sixth WTO Ministerial held that December at Hong Kong.

The consensus reached at the Ministerial and articulated in the Hong Kong Declaration of 18 December fell far short of the African Group's vision. While the declaration did contain a provision granting LDC cotton duty-free and quota-free market access in the North from the commencement of the implementation period of the Doha final agreement, the offer was a relatively hollow one.[24] Beyond this limited concession, the Declaration called for an end to export subsidies by 2006. It also re-committed WTO members to addressing cotton more ambitiously than any general formula for the reduction of agricultural support agreed during the Round. The text on the development aspect of cotton was similarly wanting. It was noteworthy mainly for a tone that bore a marked resemblance to the encouraging words deployed in the old action plans for trade and development enshrined in the General Assembly resolutions of the NIEO era. Paragraph 12 of the Declaration urged the intensification of the Director-General's work on the topic and encouraged parties to explore the possibility of establishing a mechanism to deal with foreign exchange and revenue losses. In addition to these calls for action, this paragraph prioritized further study and monitoring. It also implored the development community to disburse more cotton-specific funds and welcomed the continued processes of domestic reform in Africa that it considered to be productivity and efficiency-enhancing.[25] As the Declaration did not mention compensatory measures this idea was effectively dumped from the negotiating agenda. The action plan moving forward from Hong Kong consequently fell far short of the African Group's ambitions.

The relative paucity of efforts on the development aspect subsequent to the Sixth Ministerial was given concrete expression one year later. As part of the Director-General's responsibilities, on 15 December 2006 the Secretariat issued what it labelled an evolving working table. This table listed the sum total of 'cotton development assistance' (CDA) (WTO 2006). CDA was considered to encompass programmes, projects and activities that were both 'non-cotton specific' and 'cotton specific'. The former category enabled an array of development assistance to be categorized as CDA. Consequently, the table reported a total amount of assistance that was considerably inflated. Bona fide assistance for cotton certainly did appear. Real CDA included meetings of the Joint Integrated Technical Assistance Programme (JITAP) on cotton in Sénégal and Malawi, and the multinational programme to increase the competitiveness of African cotton spearheaded by the African Development Bank, the Union Economique et Monétaire Ouest Africaine and the FAO. However, several initiatives, programmes or projects that were listed above or below these cotton-specific forms of assistance were at best marginally related to cotton. For example, low-interest loans to the 30 African cotton-producing and trading governments under the IMF's Poverty Reduction and Growth Facility (PRGF) were included in the table. Similarly, one-off debt cancellations extended to these countries under the Multilateral Debt Relief Initiative (MDRI) launched by the G8 at the 2005 Gleneagles Summit found their way into the chart. The inclusion of PRGF funds and the MDRI inflated the total sum of proposed, announced and approved CDA to the 30 countries to nearly $6.851 billion.[26]

Forward movement on the trade aspect was limited by the slow progress of overall negotiations through 2006 and into early 2007. As such, the action on cotton that did occur was on the development aspect. On 15–16 March Director-General Pascal Lamy convened a high-level session under his mandate to bring the donor community together to coordinate current and prospective development assistance to cotton-dependent countries. The evolving working table became a focal point of the session. In his remarks, the Director-General concluded that revisions to the working table were necessary to better distinguish between cotton and non-cotton specific development assistance. He also suggested further methodological changes to ensure greater clarity while deploring the evidently wide gap between available and disbursed funds (WTO 2007a).[27] Even though the evident problems with the evolving table demonstrated that the WTO Secretariat had a limited capacity to analyse development issues beyond the realm of multilateral trade agreements, it was tasked with upgrading the document. The primary directive for the Secretariat coming out of the session actually was further work on this analytical tool. An apparently intense focus on data collection and presentation and little else underscored the virtual state of suspended animation at the WTO.

In this context the Director-General used his prerogative under paragraph 12 of the Hong Kong Declaration (WTO 2005b). He took the time at the high-level session to encourage African governments to continue their domestic reform efforts that 'aimed at enhancing productivity and efficiency' and encouraged them to deepen this process. Lamy argued that continued 'reform' – a standard euphemism for economic openness and state retrenchment – was crucial to the long-term success and viability of African cotton production. However, he failed to detail or establish the causal links between such reforms and gains in productivity or efficiency (Weston 2007: 2). Beyond pressuring these countries the Director-General noted that there was 'consensus neither on the idea of a compensation fund, nor on who would fund it' (*BRIDGES* 2007a). He implored donors to take up the suggestion made by France to explore the prospects for cotton-specific 'smoothing mechanisms'. France wanted a discussion on the ways and means to establish effective insurance schemes that individual producers could opt in to, or safety nets that would be triggered by price volatility or declines. The session adjourned without any forward movement on these issues.

An attempt to unblock overall negotiations held at Potsdam on 19 June and attended by the US, EU, India and Brazil failed to stimulate any concessions. In the wake of this meeting, calls emanated from Geneva for the US to make more significant concessions. In particular, official and non-governmental actors lobbied for the US to undertake a steeper reduction in the maximum allowable level of trade-distorting support that it could provide to its agriculture sector. The July Package had enshrined this objective and the US had previously offered to cap its ceiling on overall trade-distorting support (OTDS) at $22 billion (*BRIDGES* 2007b). The G20 proposed a sharply lower number – $12 billion – principally because the present applied level of support was much lower than the maximum the US could dole out (Khor 2007). It was estimated and later confirmed that actual US spending on trade-distorting measures was $18.9 billion in 2005, well below the $22 billion the US had tabled in negotiations (*BRIDGES* 2007c). As the US offer would only cut permissible and not real subsidies it failed to meet the additional July Package objective of reducing applied support. The Chair of the agriculture negotiations suggested that that the US should move to cap its OTDS between $13 and $16.4 billion. The US subsequently lowered its offer to $17 billion and accepted the Chair's suggestion as a basis of negotiations (*The Economist* 2007). Nonetheless, early estimates of 2006 applied levels indicated that the amount of trade-distorting support offered to US farmers that year might have fallen below $12 billion, and consequently, below the caps the G20 and the Chair had wanted to establish.

While US trade-distorting support appeared to have declined sharply in 2006, not much of this was attributable to reductions in the subsidies ruled illegal in the Upland Cotton case. The US had committed to comply with the findings of the Appellate Body in April 2005 and later eliminated its Step 2 scheme. Beyond this initial demonstration of adherence there was little evidence that the US would honour its commitment. By December 2006 measures accounting for nearly 90 per cent of the value of cotton subsidies remained unaltered or had not been eliminated (*BRIDGES* 2008a). That month Brazil formally expressed its dissatisfaction with the lack of US efforts to bring its policies into compliance with the rulings (World Bank 2007: 99). A panel was struck to take up the Brazilian complaint, and its report was publicly released in December 2007.[28] The panel found the US to be non-compliant and the US Trade Representative (USTR) subsequently initiated the appeals process, a process that was ongoing in 2009. The move to appeal came only four days after the US administration had adamantly expressed its opposition to a new US farm bill. This new package was worth $286 billion over five years and had passed by a sizeable majority in the US Senate. It did not modify the cotton programme significantly (*BRIDGES* 2008b). With the appeal ongoing, former President Bush subsequently threatened to veto the bill on the grounds that it was too expensive and would leave the US vulnerable to challenge at the WTO (*BRIDGES* 2008c). As such, the stance of the US administration was at odds with its international effort to delay subsidy reform.[29]

Confronted with palliative language from the US on subsidies and the reality of stalled negotiations, developing countries resorted in late 2007 to making collective calls for clarity and the realization of the developmental aspirations of the Doha Round. The Group of 110 formed at Hong Kong, including members of the G20, the G33,[30] ACP, African Group, LDCs, small and vulnerable economies and the C4 issued a fresh joint statement on the need for lucidity and progress in the negotiations. In a style reminiscent of the 1970s a subsequent meeting of LDC trade ministers held in Lesotho on 27–29 February 2008 culminated in the issuance of the Maseru Declaration (*BRIDGES* 2008d). This Declaration called for the end of cotton subsidies and the realization of commercially meaning-ful duty and quota free access for 97 per cent of LDC exports by the end of 2008. It also demanded that the North minimize the impact of any tariff reductions achieved in the Round on the South's trade preference margins. Maseru implored rich countries to ensure that promised levels of trade-related assistance were forthcoming. It also sought a firm guarantee that all LDCs would have access to a special safeguard mechanism to protect against import surges and enjoy heightened protection from bio-piracy. During the meeting the Ambassador of Bangladesh to the WTO told attendees that cotton represented the human face of the multilateral trade negotiations.

The poverty of the sectoral initiative

The Doha Round has bogged down into what Ha-Joon Chang has labelled a protracted 'industry for agriculture' swap (Chang 2008: 77). Even so, the harsh reality of any trade-off involving the removal of industrial pro-tections in the South and agricultural protection in the North is that poor developing countries stand to gain very little if the round is ever con-cluded. For one, World Bank researchers have projected that the transfer of income from the North to the South that would result from a likely reduction in agricultural subsidies to be barely $1 billion (Anderson & Martin 2005; Gallagher 2008: 79). If agreement is reached to fully liber-alize agricultural trade, GDP increases in the developing world are esti-mated to be on the order of 0.6 per cent. However, most of the gains from this highly unlikely scenario and from the subsidy reductions would be concentrated in fewer than a dozen countries (Wade 2005). Most returns would accrue to competitive exporters of dairy and beef such as Brazil and Argentina (Drache 2006). Many net food-importing countries would actually see their import bills rise as the global supply of key food commodities that had been bolstered by the subsidies diminished.

This impoverishing effect could be compounded by the industrial tariff cuts that the rich countries are demanding in the NAMA nego-tiations. Many developing countries maintained high tariff ceilings or bound rates when they acceded to the WTO and continue to depend upon tariffs for up to 30 per cent of their government revenues (Stiglitz & Charlton 2005: 188). While a US call for the abolition of all indus-trial tariffs by 2015 has gone largely unheeded, the prospective NAMA reductions could see tariff ceilings fall between 10 to 15 per cent. Tariffs at or below this level would undermine the revenue-raising capacity of African governments. They would also weaken the ability of policymakers there to mimic efforts other developing countries made to use tariffs as part of a sequence of measures that aimed to establish local opportunities to add value (Akyüz 2005). The policy space commodity exporters have to pursue industrial development could be diminished as a consequence. Developing countries previously ceded policy space in the Uruguay Round agreements on services and intellectual property. They also paid steep prices to unsuccessfully implement several Uruguay components, such as the agreement on sanitary and phytosanitary measures (Stiglitz & Charlton 2005: 192).

An agreement on the Singapore issues in the development round might exacerbate both of these trends. The estimated financial costs that poor countries would bear to implement new competition laws and customs procedures are high. For example, the realization of the former would impose strict limits on the ability of these countries to pursue policies that favour domestic interests. As Yao Graham detailed in early 2008, the opening of government procurement to transnational firms could also render domestic value added industries in non-LDCs such as Ghana unviable (Reilly-King & Sneyd 2008). Even if resources were available, the prospect that substantive governance changes on these issues could be achieved in African LDCs without an additionality and delivery of real resources is questionable. Absent this eventuality or an agreement to extend special and differential treatment, African governments will face an unenviable trade-off. They could embark on costly paths to reform procurement and investment policies or risk the possibility of protracted and costly trade disputes. It seems that progress on Singapore might come at the long-term cost of development.[31]

Assuming that beneficial outcomes will flow from the fulfilment of the Cotton Initiative's trade aspect, power politics, commercial interests and global governance failures have not been the only factors that have worked against its realization. Linguistic and communication problems have plagued efforts to rally South-South cooperation in support of the Initiative. Several members of the LDC and African Groups have also been inconsistent supporters. In particular, the government of Sénégal has been criticized for being overly expedient. It has toned down its support on numerous occasions when significant bilateral and regional relationships or agreements were deemed to be at stake (Diouf & Hazard 2005: 61). The C4 have also suffered from a relative lack of technical capacity to analyse and engage in multilateral trade negotiations. This dearth has been ongoing despite the existence of the non-governmental assistance programmes to scale up the C4's capacity described below in Chapter 5.

Beyond these political impediments, evidence increasingly suggests that the Initiative itself is an inadequate solution as regards the income dimension of poverty. Econometric research on the principal objective of the trade aspect has revealed this shortcoming. Most studies on the gains to be had from the removal of cotton subsidies have assumed that the supply response to less cotton coming onto the world market will not be very large, and that demand for lint will remain relatively elastic (FAO 2004b, 2004c).[32] Taking for granted that there are barriers to increasing supply and that there are plenty of substitute fibres buyers could switch their purchases to in the event of a rising cotton price, these analyses have accordingly produced low estimates of the price gains that might result from liberalization. While the numbers have varied significantly, one prominent estimate is that the elimination of subsidies would have lifted the world lint price by only 10.8 per cent between 2003 and 2007 (Benicchio 2005: 37). A few other studies on the topic have assumed global demand for cotton to be highly inelastic. The work of one of the principal development economists backing the Initiative, Louis Goreux, exemplifies this approach. His estimates of the price rise subsequent to policy changes that would shrink the global supply have been considerably higher than those that believe demand for cotton to be more price-sensitive (Goreux 2004: 6). Having surveyed this scene, Dani Rodrik (2007: 235) considers the largest credible estimated price impact of the eventual complete removal of the subsidies to be 15 per cent.

Rodrik and a growing number of other informed observers have also questioned the optimistic assumption touted by Goreux that world lint price rises would necessarily trickle down into higher farm gate prices. In their view, exogenous factors and inherent constraints would impede the ability of subsidy removal to pay off for the poorest. Regarding the former, it is possible that price rises induced by a reduction in the relative supply of the main upland cotton varieties traded internationally might not offset the impact that the declining value of the dollar relative to other major currencies has had on producers in West and Central Africa. Dollar depreciation was amongst the factors that contributed to an uptick in the dollar denominated world lint price before the global recession took hold. In early 2008, this price temporarily rose above the average evident between the 1973–74 and 1997–98 marketing seasons to 0.74 cents per pound (Weston 2007: 10).[33] Price rises associated with speculation, fuel price rises and the weakening dollar were simply not enough to compensate West Africans exporters. Their costs were tied to the appreciating euro at 655.9 CFA francs to €1, and these leaped significantly as the dollar depreciated (Baden 2004: 15). Credit, farm gate prices, ginning costs, imports from Europe, labour and transport all became relatively more expensive in dollar terms.

As such, the historic source of the CFA zone's competitive advantage in cotton production – its low costs – has come under threat. Even astute West African cotton experts such as Eric Hazard had considered the strength of this advantage to be unassailable earlier in the decade (Diouf & Hazard 2005: 61). In theory, a devaluation of the CFA franc could restore competitive advantage to the West African industry as the US debases its currency. Nevertheless, there is little historic evidence to indicate that this policy option would adequately assist poor cotton growers. A 50 per cent devaluation of the franc in 1994 simply did not raise the incomes of the poorest (Rodrik 2007: 235). Any prospective devaluation could have an ambiguous effect along the cotton chain and actually raise the costs of producers that rely on imports for inputs or implements.

Factors unique to Africa also hinder the ability of producers there to benefit from the removal of subsidies. At least one analyst has argued that with or without the elimination of the artificial oversupply, the best way to raise producer incomes on African farms is to achieve substantial productivity gains and cotton output increases (Goreux 2005: 14). This is a questionable proposition. The supply of cotton from Africa is 'not infinitely elastic' (Baden 2005: 16). At the outset, any attempt to expand conventional production intensively or extensively would necessitate additional unpaid family labour and be especially harmful for women. If the intensive option was pursued the promised productivity gains for women cultivators could well be offset. Women often do not have control over the household decision on the mixture of crops to be planted. If their husbands ascertained that intensified production was a viable way to reap greater financial rewards, men could make the apparently rational decision to devote more land to cotton. Such a choice could nullify any productivity gains that had accrued to their wives. Women who might have otherwise been able to devote increased amounts of their labour time to food crops might have to return to working similar or at best marginally fewer hours in their bigger cotton fields under these circumstances.[34] Intensified production could thereby be a precursor to extensive growth. Research on the extensive expansion of cotton has shown that it has undermined food security and fuelled environmental conflicts over water and access to fertile land across Sub-Saharan Africa down to the present (ODI 2004).

The choice to devote more attention to cotton could also be hazardous. Any reduction in the unit costs of conventional cotton or production increases achieved at the expense of women and the environment could also be offset by current global trends. Surging food prices, such as the

50 per cent increase in the world price of rice that occurred over a two-week period in early April 2008, are case in point. They underscore an emerging trade-off between the production of cotton and other inedible crops, and the food security of cash crop farmers that rely upon imported foods (Blas 2007; Beattie 2008; Beattie & Blas 2008).[35] As elevated world food prices persisted into 2010 it was clear that this trend was hurting cotton producers who were net buyers of staple foods, and was doing less harm to cotton farmers who were net sellers of staple foods. That said, while many cotton growers produce diversified harvests, most African smallholders have come to rely upon imports to meet their basic food needs. This reliance is especially acute during seasonal food shortfalls and at times of environmental stress. The most specialized cotton farmers in net food importing African LDCs where food purchases can consume up to 65 per cent of average household income continue to be most at risk. FAO Director-General Jacques Diouf warned in 2008 that the price of a basic meal in several of these countries had increased by 40 per cent over the previous year, and local prices have not abated significantly (Diouf & Severino 2008). Producers of cotton and other agricultural raw materials such as sugar might be especially hard hit by this structural trend as the world prices of the inedible cash crops that they sell have been subject to greater volatility than the world prices of the essentials they must purchase. Food price trends constitute a clear and present danger to the food security of cotton producers.

Hypothetical African supply responses to the removal of Northern subsidies are not only subject to the limits to the expansion of production described above. Any decisions taken to scale up cotton would depend upon the performance of the cotton price relative to the world prices of other commodities post-liberalization. Price rises for these crops resulting from the reduction or elimination of trade-distorting support could outperform the growth of the cotton price. Were this to occur the benefits of subsidy removal would be circumscribed. However, it is by no means certain that African farmers could readily substitute the higher paying crop for cotton or produce it competitively. The assumption that the majority of production decisions in Africa are subject to world prices is at best a stretch. Many farmers are simply unaware of these prices or are unable to ascertain them. Constraints to the supply of knowledge about world prices include deficient price information systems and the difficulties that people face accessing information and communications technology infrastructures in the remote regions where cotton is grown. Moreover, there is no guarantee that across the board agricultural liberalization will not exacerbate the relative underperformance of the cotton price

(Sumner 2007: 13). If the supply response overshoots due to scaled up production elsewhere the ostensible 'payoff' of liberalization for Africa would evaporate.

In arguing that the impact of subsidy removal on the human condition of the poorest has been over-sold, Dani Rodrik has challenged the notion that producers will reap net benefits. He has made the case that the principal beneficiaries of slightly higher border prices could well be domestic traders and intermediaries (Rodrik 2007: 235). While Goreux (2005: 84) has contested this claim, it seems to me to be plausible and worthy of further study. Absent new domestic or global regulatory frameworks to ensure an equitable distribution of the spoils, or an effective attempt to persuade the direct beneficiaries to embrace egalitarian measures, most of the minute gains to be had could be diverted to actors downstream of the farm gate. Without a transparent redistributive mechanism and monitoring system, final consumers would have to assume that cotton farmers were the real winners of price rises. If such an initiative failed to materialize many consumers would likely remain unaware of price increases upstream or how these gains were distributed due to the fact that lint makes up an ever-smaller proportion of the retail prices of textiles and clothing (Stiglitz 2006a: 87).[36] As the costs borne by final consumers would be negligible, rich urban taxpayers in the North are the only constituency set to benefit universally from the removal of subsidies alone.

Beyond the limitations of the trade aspect, the focus on scaled up development assistance and to a lesser extent compensation has also been problematic. It has largely precluded a discussion of government-led measures to shift the risks and costs associated with movements in the world price away from producers. The need to ensure the universal availability of private price risk insurance for individual growers has dominated dialogue on the topic. This focus has been ongoing despite extensive research that indicates other types of insurance might be more appropriate. For example, systems to transfer the risk of lower world cotton prices from farmers to public or quasi-public institutions that are more capable of managing that risk via forward sales could possibly be more welfare-enhancing than the establishment of a private market for insurance (Baden 2005: 98). The latter could take years to effect and is not inherently superior to public provision. Private markets could be captured by favoured private monopolies or oligopolies and these entities could impose inflated premiums just as readily as any public system could.

In several cotton-producing countries where pre-season indicative prices are offered to induce production, special funds for the purpose of mitigating price risks have been successfully established. These funds are drawn upon to ensure that cotton companies do not suffer substantial losses when world prices drop. According to Sally Baden they could in theory be maintained entirely from cash infusions when the world price outperforms expectations. Similar funds could also be established as a short-term means to mitigate the impact of exchange rate appreciation on dollar denominated cotton sales in the franc zone and in other countries where currencies are rising relative to the dollar. Burkina Faso's cotton fund performed quite successfully during the sharp price declines that occurred between 2000 and 2002. The good fortune of the Burkinabé fund indicates that this possible response to exogenous shocks is potentially a viable alternative to private schemes.

Notwithstanding the limitations of the development and trade aspects, on balance, the Sectoral Initiative is far from an instrument of pure poverty maintenance. While its objectives are clearly inadequate as regards the eradication of cotton poverty the Initiative has stimulated significant attention to cotton in the broader official and nongovernmental development communities. At the WTO, the Initiative has injected commodity questions into multilateral trade negotiations in a manner that was not seen even at the high-water mark of 'commodity power' activism during the NIEO era. It also represents the first attempt to fill the void in efforts to govern commodities since the breakdown of political, economic and technical coordination and cooperation that led the ICAs to abandon price stabilization efforts. The Initiative is also expedient at a time when exporting country officials are articulating a relatively low level of support for supply management. Evidence that buffer stocks under the cocoa and coffee agreements enabled relatively better supply management and price stability than would otherwise have been the case has simply not stimulated a drive for robust price management (Gibbon & Ponte 2005: 49).[37]

The Initiative is amenable to and reflective of the South's changing ideological approach to the governance of the global commodity trade. In advocating this measure West African states have opted to take a relatively less confrontational position on commodities than was previously the case. At UNCTAD during the 1970s these countries often argued that the establishment of buffer stock arrangements for commodities was an altogether inadequate policy option (Lavelle 2001: 36). The African Group at UNCTAD emphasized then that buffer stocks benefited rich consumers more than producers insofar as they only generated price stability. In the view they espoused at the time, this type of price management did not guarantee producers the higher relative prices that could be achieved through the indexation of world commodity prices to the import costs of commodity exporting countries. The

North-South Institute's 2008 Helleiner Visiting Fellow, Tetteh Hormeku, has attributed the downsizing of demands on commodity issues in the present era to the harmonization and alignment of output from UNCTAD's Ministerial Conferences with free market principles (Hormeku 1998). UNCTAD Ministerials were once the principal forum where counter-hegemonic appeals for remunerative commodity management were launched. From UNCTAD VIII held in 1992 at Cartagena, Ministerial documents have not been filled with language on the historic desire for commodity governance to generate rents for development. They have generally reflected a more market-based approach to governance (Taylor 2003: 409). Paragraph 21 of the 2004 São Paulo Consensus articulated at UNCTAD XI on the topic of good governance was indicative of this trend. The list of the principal impediments to 'good governance' at the domestic and international levels contained therein did not note the absence of governance for commodities (UNCTAD 2004b).[38] Orthodox thinking on the commodity question has flourished in this context.

Given the prevailing ideological climate in favour of market-based solutions to the commodity question identified by Gibbon (2007: 53), the priority treatment of cotton in the multilateral trading system and the targeting of the source of a supply glut on developmental grounds can be considered a second-best, near term solution. As one Tanzanian official remarked in 2007, 'if not at the WTO, where else?' Despite the roots of the Initiative in an appeal for freer trade, the temptation to throw the baby entirely out with the bathwater should be resisted based upon its second-best characteristic and a markedly positive externality. New advocates in the development scene have used the Initiative as a vehicle to raise public awareness of the need to alleviate and reduce poverty south of the Sahara. Through these awareness raising and capacity building actions outside of WTO auspices the Initiative has become a useful entry point for people to learn about development and poverty. As this chapter attests, the Initiative can be used as a vehicle to shed light on some of the ways that malgovernance of the world commodity trade has impoverished commodity producers and poor countries. Whether the Sectoral Initiative will have a lasting impact on poverty remains to be seen.

The end of poverty through the end of the cotton trade?

Gerry Helleiner (2006) recently reminded the development community of the difficulties involved with ascertaining the interactions between trade, trade and trade-related policies, and poverty. He encouraged policymakers to improve the poverty impact of trade and trade policies, and to focus trade attention and support on the poorest. As regards cotton, it is clear that free trade has never been tried. Trade policies have had asymmetrical impacts, and global trade governance has been more notable for its shortcomings than its successes. Participation in the global cotton trade has not been equitable, and outcomes have not been fair (Stiglitz 2006a: 62). Initiatives at the global level have thus far been inadequate to the task of ameliorating the human condition of the poorest people that are engaged in this trade. The supposed good news of the heightened attention being paid to cotton is that more work is being done by agencies and organizations to implement projects or programme assistance related directly or indirectly to developing the capacity of cotton sub-sectors south of the Sahara to trade. Unstated in many approaches to the cotton problem that view it through the lens of trade, however, is the assumption that trade in conventional cotton can pay off for poor people. This supposition is highly questionable. The cotton price is in secular decline. Poor producers are dependent upon imported foods and their remote geographic locations are a barrier to trade. Environmental externalities from production also raise the spectre of long-term impoverishment. The decline of linkages to downstream domestic producers evident during the recent period also raises questions about the type of traders African countries will become. Whether they will have the space to generate enduring gains that could be realized through the trade of cotton products with greater value added content is amongst the key questions for African cotton today.

With no end in sight to the subsidy issue, government leaders in Europe and in other donor nations are now considering the disbursement of more resources for cotton. It is possible that if increased assistance is realized, poor people in the rich world will be asked to subsidize the costs of a sector where trade horizons are bleak and the potential to eradicate poverty will be extremely limited without massive structural, institutional and ideational reforms. Consequently, the counterfactual question must be raised. What if cotton production ceased in certain producing countries and resources were devoted to the establishment of locally owned and designed alternatives, involving comprehensive education, retraining, and resettlement? Would the net financial, social and ecological benefits outweigh the costs? Given the historic relationships previously recounted, and the barriers to entry into a phase of poverty reduction that they pose, these questions cannot be dismissed out of hand. An analysis of the various economic, social and cultural costs and benefits of dispensing with cotton across the region is called for. In the meantime, African governments must be given the policy space and tools necessary to make cotton work for as many people south of the Sahara as possible.

4
Breaking the Historic Relationships in Tanzania

Tanzania's cotton conundrum will not simply be resolved through the reduction and eventual elimination of the manifold protections the US government affords King Cotton. While the economic, ethical, moral and social rationales for a prompt exorcism of this external bogeyman have been articulated and disseminated for years, and trade watchers around the globe continue to back these grounds for policy change (see Stiglitz 2006b), the removal of US subsidies cannot be a silver bullet. Those that uphold the idea that a freer cotton trade will necessarily constitute a quick win for impoverished cotton growers maintain a position that ironically parallels the rigid and determinist stance radical critics of capitalism took during the 1970s. Then, dependency theorists focused narrowly on the ways that imported governance institutions and asymmetrical linkages with the world capitalist economy fostered underdevelopment in peripheral zones such as tropical Africa (Leys 1975). Africanist scholars who underscored the inapplicability of mono-causal explanations south of the Sahara later took the unidirectional dependency perspective to task (Berman 1992).

Proponents of the liberalization cure-all would do well to revisit insights from the subsequent literature on African interactions with capitalist institutions that emphasizes the diverse outcomes of those engagements within and between countries. A brief overview of this material is also germane to the present attempt to commence a dialogue on the ways and means necessary for Tanzanian cotton growers to break out of the poverty traps that they are ensnared in.

The institutions or 'rules of the game' that shape social action and the human condition across rural Africa do not conform to the ideal-typical norm neo-classical economists espouse (North 1990: 3). Informal pre-capitalist institutions constrain so-called rational, acquisitive and

self-maximizing behaviour and ensure that the market is not everywhere and always the principal cause or determinant of social interaction. Karl Polanyi's insight that there are a great variety of fluid
and evolving institutions beyond the state and the market in which
the livelihoods of people and peoples are embedded is especially applicable to this context (Polanyi 1957: 245). Principles and norms of social
engagement and exchange are rooted in kinship, lineage and ethnic 'tra-
ditions' that have been in continual and highly differentiated processes
of adaptation and reinvention since the first African encounters with
European imperialism and colonialism. To this day, attempts that individuals in particular locales make to secure or effect a change in their
wellbeing, status and influence depend upon their ability to mobilize a
following through engaging in reciprocal exchanges (Berry 1993: 15).
Rural dwellers from diverse backgrounds consciously seek out patrons,
enter into asymmetrical or clientelistic relations of dependence and actively
work to expand their reciprocity networks. States in tropical Africa are
consequently embedded in social realities that have not only been conducive to the supremacy of patronage over policy, but have also generally
secured the primacy of personal and political 'rationalities' over economic
ones (Hyden 2006: 228).

Given these circumstances the temptation should be resisted to
make sweeping statements regarding the capacity of any given policy
initiative to alter the conditions of impersonal market exchanges or
shift governmental redistribution strategies in manners that could
foster upward mobility amongst targeted individuals and groups. As
regards cotton the perseverance of informal institutions renders blind
faith in the fundamentally pro-poor nature of the free trade solution
untenable. If the world cotton price were to rise substantially these
institutions would make it difficult if not impossible to guarantee that
increased export earnings would trickle-down into higher farm gate
prices, raise producer returns or reduce poverty. Take for example cases
where buyers and lint exporters enjoy significant market power in the
territories of several distinctive tribal or ethnic groupings. These market
players could choose to delimit any new opportunities for reciprocal
exchanges that stemmed from revenue increases to community leaders of
a favoured group, to their own kin elsewhere, or to their direct employees, agents, regulators or elected public officials. High-status farmers at the
helm of local patronage networks that govern land-use rights or power
brokers that could influence these people might also effectively undermine or nullify the anti-poverty potential of any seed cotton price rise. So
too could gatekeepers that move to augment the conditions farmers must

agree to in order to access productive tracts of land. For instance, they might demand that their clients devote more of their time to 'paying' back the favour after prices go up. This outcome would make it more difficult for poor families to sustain their production volumes and undercut their efforts to produce a quality harvest.

The fact underscored by Sara Berry that money is not the sole unit of account, store of value or medium of exchange that agriculturalists make use of also invalidates the assumption that average farmers will necessarily reap gains from a more liberal world cotton trading order. Growers that sought to take advantage of any higher price that did materialize might enter into many reciprocal social obligations to boost their access to labour and their outputs. In so doing it is entirely possible that these people could overextend themselves into the future.[1] Taken together this cultural milieu and the evident polarization of incomes and wealth across rural Africa over the past three decades militate against the development effectiveness of moves to redirect resources away from subsidies and induce a global supply shortage. This setting similarly complicates the task of those in the international development community that have taken a more nuanced view of the cotton problem. These persons actively seek to secure an additionality and delivery of real resources to the farmers most in need through the auspices of the World Trade Organization (WTO). Moreover, it makes any attempt to map the possibilities for poverty-free cotton production into a daunting task loaded with potential pitfalls.

From the outset then, the limited and incomplete nature of this case study of Tanzania must be recognized and stressed. The story of the governance and market reforms that need to be undertaken to break the factors that have generated and maintained poverty in the past is necessarily partial and contingent. What follows is a prelude to a conversation that has no pretension to finality. Nor does it mean, as Professor Sam Wangwe put it during one memorable interview, having all the answers at a time when evidence of universally successful anti-poverty strategies is in short supply and information asymmetries are abundant. This chapter is a stocktaking exercise and determinedly not an attempt to assert the superior intellectual rigour of any specific approach to understanding and acting upon poverty. It draws upon income-based understandings, insights from the capability deprivation school, the concept of social exclusion, participatory poverty research and elite and on-farm interviews.

The narrative commences with a presentation of elite perspectives on the changes necessary to fight poverty. Here, standpoints on cotton-specific reforms are integrated with ideas on how to break broader relationships that have fuelled the relative and absolute impoverishment of Tanzania's cotton growers. Next, five dynamic factors that bear upon progress are detailed.[2] The final section briefly recounts several of the unique tales rural people imparted to me about cotton, their struggles to make it pay and what they have told other researchers about the direction of poverty trends.

Elite perspectives on poverty eradication

Similar to other highly political concepts such as democracy, freedom, or justice, there is no universally accepted definition of poverty. The only non-controversial statement that can be made on the subject is that the condition entails a level of deprivation relative to an average standard of living. All attempts to define the concept in more concrete terms contain arbitrary or subjective elements (Stewart et al. 2007). Despite the contest over the meaning of the term and the fact that different understandings belie efforts to arrive at objective ways of knowing about deprivation, donor, 'partner', academic and civil society researchers continue to devote considerable resources to refining the concept and improving measures of poverty. This focus has been fruitful insofar as it has led to the emergence of the view that poverty has multiple dimensions.

As early as 2001, a survey of the available literature found that poverty researchers were no longer solely concentrating on incomes or on explicating the bundles of goods and services necessary for people to meet their 'basic needs' (Sandbrook 1982). The report noted that many had been busy operationalizing Amartya Sen's (1999) theory that poverty entails the deprivation of capabilities. They had also detailed the exclusions, powerlessness, stigmas and vulnerabilities associated with poverty in particular contexts through ascertaining what people themselves understood 'poverty' to entail (Narayan 1997). Nonetheless, Frances Stewart and other contributors to the latest comprehensive study on poverty research have concluded that the field continues to be dominated by income-related topics including the absence of assets, lack of control over earnings or insurance failures. Mainstream approaches to measuring poverty have also continued to rely excessively upon income-based indicators. While powerful researchers might assert the need for 'human and physical asset increases and safety nets' (Lipton and Ravallion 1993) or claim that 'economic, environmental and social sustainability' is an overarching imperative (OECD 2006: 89), most high-level output on the topic exudes the idea that poverty is measurable in terms of incomes. For example, the targets detailed in the UN Millennium Declaration of September 2000 now known as the Millennium Development Goals (MDGs) do not exude a broader view. The goal to 'eradicate extreme poverty and hunger' by 2015 has been delimited to an attempt to halve both the incidence of absolute income poverty (defined by the World Bank at the time as one US dollar per day at purchasing power parity) and the proportion of people that suffer from hunger, and also to create more income-earning opportunities for the poor. Other millennium targets did not fall under the heading of poverty eradication. These goals, including the achievement of universal primary education, the promotion of gender equality and empowerment, the reduction of child mortality, and moves to combat disease, realize environmental sustainability and establish a global partnership for development, were presented separately. As such, the goals that were to drive development policymakers distinguished between poverty, defined in monetary terms, and 'social development'.

Despite the definitional conflict, the competing approaches to measurement and the vast available literature on poverty at the country-level, there is one certainty. The last thing that Tanzanian cotton growers need is another study authored by an official or independent national, expatriate or foreign-based researcher with pretensions to the objective treatment and scientific measurement of poverty challenges. An attempt to measure precisely which factors are operative in particular locales across a cotton zone that covers 42 districts and 13 of the country's 21 regions would be a costly, if not an impossible exercise. Such an approach would be loaded with politics and potentially not add very much to what is already known about the relationships that keep people poor.

My interviewees in Dar es Salaam and across the sub-sector highlighted a diverse array of factors that have impoverished people and underscored the point that poverty does not bear upon all cotton farmers, households or communities in the same ways. Their words and my own desk research enabled the construction of the list of feasible interventions presented below that I feel are necessary to overcome multidimensional cotton poverty. This account of the possible paths to augmenting incomes and capabilities and reducing social exclusion is followed by a discussion of five dynamic issues that could make or break the agenda to alleviate, reduce and eradicate cotton poverty. It is possible that domestic governance reform and resource mobilization, the management and distribution of donor resources, biotechnology policy, foreign direct investment and ongoing reliance upon lint exports could facilitate an enabling environment for the improvement of livelihoods. It is equally the case that the failure to effect changes in any of these areas could undermine or preclude the interventions and redistributive measures necessary to improve conditions on the farm and in rural communities for the four in ten

Tanzanians who are directly or indirectly dependent on cotton (Forum-Coton 2004).[3] To reiterate, what follows is a prescriptive list derived from the perspectives my interviewees articulated. This re-presentation has been written in prescriptive language to reflect the opinions that various individuals conveyed to me. While I agree with many of the positions detailed below, this list should not be characterized as being a list of 'my' prescriptions *per se* – what follows is simply a compendium of possible ways forward.

Domestic governance reform and resource mobilization

Innovations in input systems could raise productivity and reduce aspects of poverty related to low incomes and capability deprivation. For example, a new financial mechanism to alleviate chronic producer indebtedness and a targeted fund to make seasonal credit and investment capital more readily available would give producers reasons to choose to stick with cotton (Poulton 2006).[4] An overhaul of the seed distribution system is also desirable to ensure that ginners have adequate incentives to set aside the best seeds for planting and that farmers receive these quality seeds after they have been subjected to a safe de-linting process. As with the pesticide distribution system, officials at the village, ward and district levels, buyers, their agents and Tanzanian Cotton Board (TCB) employees must all have incentives to ensure that gatekeepers at distribution points stop supplying their relations and friends with 'free' seeds and that seeds are distributed in an equitable manner.

It is also important to weigh the relative costs and benefits of the sub-sector's reliance upon a local producer and an international trader that both supplied chemical pesticides under contract to the now defunct Cotton Development Fund (CDF). The trader had sourced pesticides manufactured in China, and in the interest of producer and environmental health, it is worth evaluating alternatives. For example, the prospects for stimulating the creation of an employment-generating upstream industry to produce and supply botanical pesticides to combat aphids, cotton stainers, jassids and pink bollworms should be investigated. Organic fungicides to take out Fusarium wilt should also be studied. Organic fertilizer usage – depending upon the source, currently as low as 1–2 per cent in certain districts or as high as 30 per cent nationally – could also be promoted in a way that enhances land husbandry practices.

Work must also be done to guarantee that animal traction teams are managed in a more environmentally sustainable manner. The will of women and men to continue to cultivate and weed in teams with hand hoes needs to be assessed, and measures put in place to ensure that the cultural practice of collective labour is respected, remunerative and productive. Despite the World Bank's (2007: 151) attempt to characterize subsidies as not very 'market smart', they might prove to be 'poverty smart' if they are deployed and managed effectively in the service of any of the above reforms. Disbursements to mitigate prices and costs associated with the adoption of these advances will be essential to their viable realization.

Additional resources for research, extension and quality, and upgraded management systems in these areas could also reduce poverty. The seed variety most commonly used in the WCGA, UK 91, is nearing the end of its viability.[5] Ukiriguru, the agricultural research centre for the lake zone responsible for developing the seed, remains grossly underfunded. The research and development of new seed varieties that enable production in new districts could also be desirable if resources are simultaneously devoted to the development of contingency plans and safety nets to mitigate the risks and threats associated with the pursuit of cotton cultivation in new districts.

Cotton-specific extension services must also be developed to replace a system that many consider to have broken down. The extension services available on bioRe Tanzania's certified organic cotton project in Meatu, Shinyanga detailed below in Chapter 6 could serve as a model. bioRe farmers enjoy services that contrast sharply with the opportunities for technical advice or training currently available to Meatu's conventional growers. Whatever form extension takes, the design, implementation, monitoring and evaluation of farmer training and productivity enhancement must be participatory and involve close collaboration with the private sector. Buyers have an interest in securing consistently higher output levels of quality seed cotton. An extension system that imparts techniques to reduce bacterial blight and other hazards and also helps households to redistribute responsibilities and incomes from cotton in ways that enhance productivity is also a win for ginners. A new focus on extension could also remedy the seed cotton contamination issue that has plagued the sector if it raises the capacity of families to ensure that they can safely produce a quality crop. Complementary efforts to introduce collection sacks made from cotton or other natural fibres and discontinue the use of plastic bags could also augment quality and create jobs.

The marketing infrastructure also needs to be improved. After intense rainy seasons minor roads in the cotton zone can become impassable to donkey or ox carts. At those times many small growers must head load their crop to market and navigate around mosquito-infested washouts.

District cotton task forces could engage with ward and village level officials to identify potential problem areas and avert the costly downloading of infrastructure failures onto cash-poor and time-constrained farmers. Price information systems should also be developed to enable members of primary societies, buying agents, village executives and more marginalized producers to compare the prices on offer in their ward with prices offered elsewhere in their district or beyond. The cotton inspector could compile a list of relevant cell numbers, make cell phones and credit available and inform sellers in remote locations about this service. Storage facilities are also often inadequate to the task of keeping cotton either clean or dry. Inspectors have found many of these buildings to be in technical violation of TCB standards year-on-year. Ginners could reap the reward of higher quality lint if their agents are asked to assess and upgrade cotton storehouses.

At the market, ginners must also ensure that their agents have enough cash on hand to make spot payments to producers when prices are high. An emergency stabilization fund, public insurance programme or other comprehensive price risk management strategy should also be designed and established through multistakeholder dialogue with an eye towards protecting the poorest. This mechanism could be triggered whenever a global supply glut results in lower absolute prices at the farm gate, or when other exogenous shocks such as the rising cost of imported foods, essentials or currency depreciation fuel a reduction in the purchasing power of seed cotton earnings. If the TCB resumed the issuance of indicative prices these could be indexed to changes in the prices of a basket of goods farmers themselves deem essential to 'leading a life that they value', and disseminated to the development community in Dar es Salaam and beyond. Moreover, if the Board were able to command the resources and capacity to launch regulated auctions to replace the current system of private buying posts it could introduce the indexed price as the floor price for all seed cotton purchases in a given season and advance the poverty eradication agenda. The status quo has proven to be ineffective due to the persistence of informal price-fixing agreements, the underservicing of remote districts and the practice of irregular staffing. The latter generates inefficiencies when farmers arrive at closed buying posts on days that they have allocated to the task. A move to embrace auctions would reduce buyer margins and might not be necessary if the yield and quality-enhancing interventions prescribed above take hold in a manner that enables producers to keep their costs low and raises their incomes.[6] Even so, indexation could help all players in the sector to better comprehend and determine what exactly the 'fair market price' for seed cotton is each season.

Buyers could also move collectively or be induced to embrace an out-grower model. The Tanzania Gatsby Trust, a registered charitable trust for the relief of poverty and the advancement of education, has recently backed the creation of an outgrower scheme in Sengerema District. This project will also involve the introduction of demonstration farms through-out Mwanza Region, and has been undertaken with the intention of enabling ginners and farmers to weigh the merits of contract farming (*The East African* 2008a). The Board must ensure that existing and new investors have an incentive to embrace this model and provide credit and inputs on credit. One way that it could do this would be to inform donors, other development finance agencies and domestic banks about the benefits that could be reaped if greater levels of pre-season credit were made available. Ginners might also have an incentive to move into the business of working closely with farmers to ensure sustainable, high qual-ity harvests if the Board were to simplify its complex licensing and buying application processes and these reforms freed up human resources and funds.

Private buyers also want to see that the taxes they pay to the districts are used for investments that have visibly positive impacts on their oper-ations. Members of district level cotton task forces could be empowered to canvass the opinions of buyers, village executives and farmers as regards the allocation of district disbursements. An annual survey on this topic could draw the attention of district officials to the concerns of major tax-payers and residents and complement the new mandate task forces have under the Board's *Corporate Strategic Plan* (CSP) to set and realize pro-duction targets. Perhaps the most important reform necessary to free up investor resources for upstream support is the eradication of the *machinga* problem. The buying licenses of any ginners that are found after invest-igation to have either purchased cotton from unauthorized traders, tacitly supported the activities of illicit traders, or covertly hired agents to de-stabilize outgrower schemes must be revoked. Success or failure on this imperative will ultimately depend upon developments related to factors discussed below.

Several other reforms are necessary to ensure that the levels of social exclusion within households and communities across the sub-sector are reduced. An extensive process of consultation should be launched to determine the appropriate and most viable scale for cotton produc-tion. Given the diversity of farm sizes and life ways across Mwanza, Shinyanga and the other cotton-producing regions it is desirable that context-specific models are developed. The Board has vowed to help commercial farmers secure formal title to more land, but it is not clear that a greater commercial orientation is in the interest of the 95 per cent or more of cotton growers that could be described as family farmers. Many of these smallholders do not have formal title. This fact has limited their access to formal credit and insurance. It has also left them vulnerable to any move to expropriate vast tracts for new exporters or foreign investors seeking to establish large-scale farms. Customary tenure must be recognized, formalized and documented so that producers can decide for themselves what is to be done with their farms. In districts where commercial scale production is pursued, mechanisms to smooth farmer exit such as relocation and retraining assistance should be made available.

To give farmers a voice in the future direction of the industry, funds could be disbursed to back the expansion of the Tanzania Cotton Growers Association into a representative and effective national organization. The problems that accompanied donor-inspired reforms and the provision of donor credit to new entrants in the sector who were subsequently unwilling or unable to assist their cotton 'suppliers' could be redressed if domestic players ante up in support of producer organization. Women must be prominent participants in this association and also in all village, ward, district, regional and national multistakeholder and political dialogues. Their input and empowerment is fundamental to the realization of more socially inclusive outcomes. The rights of women, children and the aged to share in the benefits of cotton production and to participate in a value chain that is free from skewed intra-household or community distributions of incomes from cotton must be enshrined at the heart of all TCB and private sector operations. More broadly, the access that marginalized people have to secondary education, healthcare and social and veterinary services needs to be enhanced to improve their capabilities and functionings.[7] Few elite interviewees spoke to me about what happens to elderly producers when they can no longer work their fields or about the burden of elder care. Unawareness of this problem and inaction on exclusions from services more generally must change if policymakers are to substantively address the time dimension of poverty.

Environmental management is also a significant poverty challenge. A specific drought management strategy should be put in place. When an event of similar extent or duration to the 2006 drought occurs in the future, cotton farmers should not have to rely exclusively upon the ad hoc interventions of donors or socially-conscious buyers. The benefits of a greater role for irrigation must also not be assumed. An effort to scale up the percentage of cotton under irrigation could potentially divert investments in the realization of pro-poor and sustainable input systems or infrastructure upgrades. While the net return on irrigated hectares could be substantially larger than on rainfed plots, benefits from irrigation might accrue in a highly concentrated manner. These could also be short-lived if irrigation is pursued on large-scale farms, intensive techniques are used and despoilment ensues. This is not an appropriate technology for poverty reduction if farmers could otherwise reap gains from a new input system that reduced health hazards, preserved soils, improved the safety of animal feed and fodder and possibly put many growers on a path toward organic or other high value certifications.

The challenges of potable water availability, water source maintenance, safe sewerage, suitable drainage and solid waste disposal are also omnipresent in the cotton zone. All stakeholders must recognize that these problems bear upon productivity and yields over the short, medium and long terms and that the status quo is a considerable barrier to output growth. To borrow applicable language from the WTO, inter-generational equity and the interventions necessary to ensure it must be embraced 'specifically, expeditiously and ambitiously'. Anti-siltation and erosion programmes, biodiversity awareness and preservation plans, moves to minimize the impact of extensification, forest landscape restoration and the formulation and implementation of sustainable energy systems are particularly important points of departure (Moseley 2005).

A range of factors previously elaborated could confound attempts to realize these pro-poor prescriptions or reduce their effectiveness. Production and yield increases or cost declines elsewhere, heightened global demand for synthetics and the persistence of subsidy regimes would not be helpful (TCB 2007). Adverse market trends would be even more severe if cheap textile and apparel imports – the corollary of Asian demand for Tanzanian lint – prevent a concerted attempt to move downstream where as much as 90 per cent of the income and employment is generated along the chain.[8] Beyond these and the other ideational and institutional conflicts at the global level discussed above the issues of domestic governance reform and resource mobilization are the first of five important determinants of the potential for poverty reduction. As one of my respondents put it, failure to improve governance or to mobilize resources will result in cotton continuing to be 'grown by default and not by design'.

To its detriment the Board's CSP downplayed these variables and poverty issues more generally. In an overly economistic leap of faith the plan's authors claimed that a doubling of lint output from the 700,000 bales processed and sold during the 2005–06 bumper season and a doubling of seed cotton yields from 750 kilos per hectare would go a long way towards the 'eradication of poverty and the improvement of human welfare'.[9] To achieve these ambitious targets by 2009–10, the CSP proposed the extensive expansion of cultivation and articulated a plan to raise the productivity of the 350,000–500,000 farmers that register to grow cotton annually.[10] While concepts such as 'incomes' and 'quality of life' found their way into the list of key performance indicators included in the document's logical framework, the Board's primary objective over the period of the plan remains the 'sustainability of cotton'. Poverty eradication and livelihood improvement also failed to be listed amongst the plan's 'core values' of wealth creation, downstream linkages, sustainability of cotton, professionalism and innovation. As such it seems that the CSP tacitly equated output and productivity growth with poverty reduction. The plan simply did not address the capacity of the TCB to incorporate the poverty targets articulated in the National Strategy for Growth and Reduction of Poverty known as MKUKUTA into its operations and the activities of its agents, nor its ability to ensure that private buyers or their sub-contractors are furthering these goals.[11]

The TCB's limited governance reform plans and the persistence of low-level corruption could hold back efforts to improve regulatory performance, reduce rent-seeking opportunities and foster social inclusion and equity. The Board has vowed to augment its corporate governance through instituting financial controls and systematizing the allocation of its financial and physical resources. However, it has not developed concrete objectives for governance improvements in the private sector. This task has been effectively off-loaded to external agencies. Even though cotton inspectors have been tasked with enforcing the Cotton Act it is not uncommon for regulators to characterize the scale-rigging issue or anti-competitive practices respectively as falling under the purview of the Tanzania Bureau of Standards or of the embryonic Fair Competition Commission. The delimitation of governance reform plans to in-house affairs is also problematic given the Board's efforts to authorize district task forces to formulate, implement and supervise cotton production plans.

Much like the 'bottom-up' approach enshrined in the Agricultural Sector Development Programme, the extent to which new priorities for cotton articulated at the district level reflect the interests of the majority of producers will depend upon the ability of women and other marginalized farmers to exercise voice. If district task forces exclusively consult relatively rich male producers it is possible that the distribution of any benefits from new initiatives will be skewed towards people who were able to attain their present status and wealth through the old male-dominated cooperative system. The TCB has not detailed how task force members will be compelled to embrace the principle and practice of transparency or be held to account for their actions. As such, it has effectively relieved itself of responsibility for any future governance failures that impede production targets or entrench inequalities.

That being said, inadequate resource mobilization could be at the root of the persistence of corruption in the cotton zone.[12] The TCB does not command the resources necessary to ensure that the principals and agents of corruption are subject to greater scrutiny or to legal challenge. Under the Medium-Term Expenditure Framework (MTEF) the Board is now dependent upon annual budgetary disbursements. If the TCB were to make an autonomous effort to augment its regulatory capacity in this area the attempt might be easily frustrated. The Board no longer can impose levies on exporters or access external resources. It simply has no control over the relevant levers of domestic resource mobilization (Culpeper 2006; UNCTAD 2007). This lack of financial independence compounds a problem numerous interviewees identified regarding its authority to enforce the Cotton Act.[13] There is talk that certain ginners and traders have cultivated relationships with officials at the Ministry of Agriculture. These figures have attempted to ensure that they have influential advocates on their side in the event that the Board actually moved to impose penalties for violations of the act or to revoke buying licenses. It is also possible that corruption has been a factor underlying both the evident failure to enforce buying post standards and the persistence of tax evasion. Deficient capacities or willingness to engage this issue will only be bested after a concerted attempt to impose powerful sanctions delegitimize the practice and eliminate the mechanisms that have maintained it. Beyond the realm of cultural economy variables one of the possible mechanisms that have upheld corrupt practices is the aid regime.

Donor resources

Taken together, flows of donor resources, approaches to aid management and the knock-on effects of development assistance are the second dynamic factor that bears upon poverty outcomes. In the mainstream view, donors need to scale-up investments in agricultural development to rectify a long-term decline in the share of donor aid to the sector (UNMP 2005; Sachs 2005; Wolf 2005a, 2005b; Jayne et al. 2006). Coupled with the realization of greater export market opportunities and a lower debt burden, aid campaigners have argued that more disbursements could enable public investments in infrastructure, input systems, institutional capacity and education programmes. External resources could boost productivity and also alleviate dislocations associated with deagrarianization such as labour redundancies, deskilling, forced migrations and a lack of rural employment opportunities. Despite the fact that over 80 per cent of Tanzania's 40 million citizens work in the sector aid advocates have underscored the point that Official Development Assistance (ODA) to agriculture constituted less than 6 per cent of total ODA disbursed during the 2000–05 period (World Bank 2008).[14] Even so, Tanzania has received net levels of assistance consistently two to three times above the pan-African average. As such, the dollar value of donor funds that have flowed to the sector is high relative to the flows agricultural sectors in other cotton-producing countries have received (Helleiner et al. 1995).

Since Professor Gerry Helleiner led an independent review of the state of the aid regime 15 years ago the counterintuitive proposition that Tanzania might be receiving too much aid has been aired. Researchers have probed the connections between increased levels of assistance and welfare enhancement. The idea that the opportunity cost of chronic 'aid dependency' could be foregone corruption eradication or perpetual poverty has become increasingly salient (Sneyd 2007; Bendaña 2008).[15] For example, Brian Cooksey (2003) concluded that smallholders are often caught between a 'government/donor rock and a private sector/market economy hard place' with few benefits or incentives, but plenty of poverty. His work did much to advance the notion that aid can be an instrument that differentially empowers the policy elite and wealthy farmers. Critics of aid have also argued that a surge of new funds could increase the value of the shilling – a Dutch Disease-like effect – and an initial investigation of the impacts that more aid could have on Tanzania's real effective exchange rate has been conducted (Li & Rowe 2007). This eventuality would enable the well connected to take advantage of the opportunities a higher currency valuation affords to consume more abroad while raising the possibility that traditional exports might be priced out of global markets. The resulting cost-cutting imperative would fall squarely on the shoulders of smallholders (Easterly 2006). Aid can also play an ambiguous role as regards the balance of payments 'constraint' on poverty reduction policy (Kanaan 2000). Tanzania's current account deficit was recently as high as $530 million and it is important to consider the contribution that aid disbursements make to the import bill and ask whether aid is capable of generating an improved balance of payments position.

Importantly, the idea that Tanzania is somewhat of a donor 'darling' no longer holds sway. A 2004 controversy over the purchase of a $40 million presidential jet, revelations that came to light in 2007 concerning a questionable deal to procure an air traffic control system from BAE Systems, and the exposure of a series of fraudulent draws on the Bank of

Tanzania's external payments arrears account have reduced the Government's reputational capital.[16] In August 2008 several of Tanzania's 'development partners' took issue with the Government's financial management systems when they sat down to discuss the release of funds for the 2008–09 budget. According to media speculation, several members of the Group of 14 donors that provide general budgetary support to the treasury under the country's Joint Assistance Strategy refused to disburse between $677 million and $2 billion until they were informed of the specific measures that would be taken to improve oversight and prevent fraud (*The East African* 2008b). While talks with these partners eventually succeeded in releasing the funds, donor contributions to the 2008–09 budget dropped to 34 per cent from 42 per cent the previous year. Officials portrayed the reduced donor role in the budget of 7.22 trillion shillings or $6220 million at market exchange rates as evidence of greater 'self-reliance'.

It is far from certain that donors will deliver more euros or dollars moving forward. The prospect that members of the Group of 8 might not honour the commitment they articulated at the 2005 Gleneagles Summit to double aid flows to Africa by 2010 through directing an extra $25 billion per year to the continent is a real one (Tomlinson 2008: 200). As the global recession took hold in October 2008 donor governments increasingly put public funds on the line to recapitalize failing banks and other financial institutions, build trust, augment the unsuccessful efforts their central banks had made to unblock inter-bank lending and prevent a freeze in commercial and retail lending. With global stock markets tanking, in his capacity as Chair of the African Union that year President Kikwete admonished donors to not let their costly new policies to rescue the financial sector exert downward pressure on aid budgets. If Kikwete's fears are realized Tanzania might have to make due with current levels of support. Even if new funds were to materialize they might not work equally well for all of the innovations necessary for poverty eradication (Cooksey 2004; Sagasti et al. 2005).

Alternatively, Tanzania could turn increasingly to China for the provision of official credit or to any private creditors at the global level that continue to be willing to lend. These options could in some ways depart from the spirit of the Accra Agenda for Action (AAA) reached at the Third High Level Forum on Aid Effectiveness in September 2008. At Accra, members of the OECD-DAC embraced a broader conception of 'country ownership'. Donors now consider the presence of a national development plan that includes a poverty reduction strategy and is linked to the country's MTEF for budgetary planning to be a necessary condition for the realization of ownership, but not a sufficient one (OECD 2005).[17] In official terms, better ownership entails broadening country-level dialogues on aid and the inclusion of civil society inputs. It involves moves to untie aid and make it more predictable, and also the adoption of a new approach to make certain that any conditions attached to disbursements are transparent, public and drawn from the recipient's own plans (OECD 2008). While implementation might fall short of aspirations, there are significant risks associated with greater reliance upon China or on other creditors who are not aligned with this agenda.

Dependence on non-aligned donors could expose Tanzania to *quid pro quos* with high social, environmental and development costs, and undermine its commitment to debt sustainability. On the former, the Government might be asked to support prospective Chinese investors through offering up cheap resource concessions or reducing the royalties or taxes these firms would otherwise have to pay. It might also feel compelled to create regulatory exemptions or to provide matching investments that do not square with farmer needs. Even if more funds with fewer strings were forthcoming the same actors who have been able to exploit the aid system in the past might secure disproportionate shares of the spoils to be had from new entrants in the development finance scene. The advent of South-South development assistance is not inherently 'win-win'.

Biotechnology

The as yet unresolved place of biotechnology is a further dynamic factor that could affect the realization of pro-poor outcomes. The Government of Tanzania has adopted a precautionary approach to the creation of legal and regulatory frameworks to govern the introduction of biotech crops. This non-committal stance has stemmed from considerable political opposition. Resistance first burst onto the national scene after a series of newspaper articles claimed that varieties of genetically engineered maize had entered the national food supply via the provision of food aid. A coalition of environmental advocates opposed to the ad hoc adoption of genetically modified (GM) organisms was subsequently solidified. The alliance gelled after it came to light that the Tropical Pesticides Research Institute – an organization established by the Government in 1979 to test agrichemicals – had conducted field trials of GM tobacco, and reports surfaced that GM cotton trials were immanent (Hosea et al. 2004; *The East African* 2005). The Ministry of Agriculture had approved the former experiment under the authority of the Plant Protection Act 1997. Trials were halted only after the attention of officials in the Vice President's Office and other ministries was drawn to sections of the Environmental Management Act 2004 (URT 2005a) regarding the management and mitigation of risks associated with the release of living modified organisms. The fact that a national biosafety protocol was not yet in place was also a key rationale behind the standstill.

To inform the ensuing debate between the figures at the Ministry of Agriculture that were supportive of the new technologies and other concerned officials and members of the public, a National Biotechnology Advisory Committee was struck. The Ministry of Higher Education, Science and Technology convened this committee, and a secretariat was subsequently set up at the Commission for Science and Technology. Efforts are ongoing to realize a national biotechnology policy. At the time of writing the National Biosafety Protocol remains in draft form (URT 2005b). This slow pace has led to accusations that the Government is confused. It has also made several environmental activists feel as if they have been left 'in darkness'. Officials have played up the strategic dimension of this pace. They claim that it can counter the power of the biotechnology lobby. Be that as it may, the idea that people have been kept in the dark for their own safety must be balanced with the acknowledgement that slow forward movement could also reflect deficient human resource capacities. This speed is also potentially conducive to rent-seeking. Powerful lobbyists have maintained an interest in the Tanzanian situation and its resolution in their favour, and at least one bilateral has splashed money around to advance their cause.[18]

The balance of present evidence suggests that there are more questions than answers about the exact impacts the introduction of *Bacillus Thuringiensis* (Bt) cotton could have on poverty.[19] Contrary to the OECD's (2006) positive appraisal and the World Bank's (2007: 158) characterization of transgenic, insect resistant cotton as a 'win-win-win' for yields, farmer profits and producer health, it is likely that the institutional context in Tanzania is inauspicious.[20] Recent empirical work on the once-heralded Bt cotton experiment on the Makhathini Flats of KwaZulu-Natal, South Africa has documented a precipitous decline in cotton cultivation (Gouse et al. 2008). Yields there have improved considerably and the Bt crop outperforms traditional varieties during its rainy season growth phase. However, the availability of credit for seeds, fertilizer, animal traction hire and seasonal labour has been deficient. Inadequate training and extension services have also fuelled a switch out of cotton and into less labour intensive and costly crops. This outcome has occurred in a country that commands a greater level of resources to devote to the provision of agricultural credit and services than Tanzania.

Proponents of trials and the thoroughgoing adoption of Bt cotton are aware of the mixed results. They are nonetheless excited by the prospect that it could overcome pink bollworm infestations. Bt cotton could also enable the Board to allow production in a zone south of the Rufiji River that has been used as a buffer against pest outbreaks in Malawi and Mozambique. Other hypothetical rationales for change include the need to overcome the troubles many smallholders have financing two or more sprays per season, the widespread practice of barefoot spraying and the recycling of pesticide bottles for household use. There is also a risk that other lint exporting countries that adopt the technology will realize lower production costs and undercut Tanzania's apparent comparative advantage.

On the other hand, the challenges Bt cotton could pose to improvements on the farm are manifold. First and foremost are questions regarding the contract terms necessary to ensure that proprietary seed technologies are not impoverishing. Who will determine the criteria for awarding tenders? What processes will be put in place to guarantee that the Government and farmers are not beholden to a single overseas supplier for an extended period of time? Are growers going to be asked to pay royalties and bear the full costs of the programme year-on-year, or will they be able to assert what Craig Borowiak (2004) has described as their subjective farmers' rights to inform and moderate the process? Can foreign seed dependency be transformed into viable technology transfer and local ownership if resources for research and development are made available to adapt the technology to local conditions? To enable this work, will rich countries waive the obligations to honour patents Tanzania and other African WTO members acceded to under the Trade-Related Aspects of Intellectual Property Rights (TRIPS) agreement? Alternatively, will Tanzania have to pursue the patent infringement strategies that Japan and Germany deployed during their quests to industrialize (Chang 2008: 122)?

Moreover, there are numerous environmental red flags and uncertainties surrounding the possible impacts forward and backward from smallholdings (GRAIN 2003). Why has the debate over the role of Bt cotton in integrated pest management been so fierce (Hillocks 2005; Williamson et al. 2005)? Is it possible that this technology will fuel pest-resistance over the long-term or lead to the resurgence of non-targeted pests? If these latter eventualities come to pass and farmers increasingly pursue cotton monoculture, as Michael Lipton (2005) has wondered, would it even be possible for authorities to assure an effective emergency response? The upstream and downstream effects are also far from clear. Are job losses in the seed de-linting and distribution business going to be offset by new hires for research or extension, and if so, who will foot the bill? Given that the Board has prioritized scaling up the production of cottonseed oil and derivative products such as oilseed cake animal feed, what are the possible financial ramifications of Bt dependence for producers and health implications for consumers (Witt et al. 2006)? Is it possible that the introduction of biotechnology will rule out the creation of a potentially lucrative value added industry to produce botanical pesticides or come at the cost of underinvestment in the country's proven organic farming capacity? Finally, the market for GM lint is by no means a sure bet. Dunavant, a now-defunct US-based international cotton merchant, expressed reservations about its physical properties. This firm noted that these properties are deficient relative to the characteristics of particular national origins that spinners continue to value highly.

Foreign direct investment

Similarly, there are considerable uncertainties over the role that foreign direct investment (FDI) could play in breaking the factors that impoverish conventional cotton growers. The capability of FDI to spread better working conditions or ratchet up environmental standards often touted in the pages of *The Economist* simply cannot be applied with broad brushstrokes to Tanzania's agricultural sector. As a renewed scramble for African resources has taken hold during the recent commodity boom and privatizations have opened up spaces for new investments in services, FDI inflows across the continent have increased considerably (UNCTAD 2008). However, the $36 billion USD in new investments Africa received in 2006 was not enough to alter the historic downward trend in its share of global FDI flows. Its share that year fell to 2.7 per cent (Bond 2005). Investments in fixed capital formation were concentrated in Nigeria, Angola and South Africa. Tanzania received only $377 million in FDI inflows in 2006, a figure that increased its total FDI stock to roughly $6.1 billion. Many of Tanzania's post-liberalization investments have been concentrated in manufacturing and in the new mining and tourism sectors. Agriculture accounted for less than 5 per cent of the total value of investments approved by the Tanzania Investment Centre (TIC) from September 1990 through June 2004 (URT 2004a). Incentives such as zero-rated duties on capital goods imports and value added tax (VAT) deferments were insufficient attractants.

Supply-side constraints have continued to impede the TIC's efforts to attract agricultural investors while it has become increasingly apparent that FDI could have a marginal or ambiguous effect on rural poverty. Researchers at Dar es Salaam's Economic and Social Research Foundation have cited deficient transportation infrastructure, low levels of human capital, the high costs of setting up irrigation systems, lengthy gestation periods, tenure formalization delays, clientelism and petty corruption as principal determinants of the shortfall.

Another observer whose recent work lauded FDI as a force to offset domestic investment gaps, transfer technology and improve the productivity of the poor has argued audaciously that smallholder agriculture itself is the source of the bleak investment climate (Msuya 2007). This study came within a hair's breadth of blaming rural people for the outcomes of the institutional failures they have been subjected to. It entirely failed to question the appropriateness of the technologies deployed or to discuss job creation. Nevertheless, it did highlight the merits of several existing outgrower schemes. Such investments in the sugar and tea sub-sectors have stimulated the extensive growth of production volumes, raised yields and improved quality. This model has not as yet spilled over or had a demonstration effect on investors in the conventional cotton business.[21]

It is also possible that the large-scale investments of this type could perversely reduce the domestic food supply and raise sustenance costs for agriculturalists that grow inedible export crops moving forward. Jacques Diouf, Director-General of the Food and Agriculture Organization, has warned of an incipient 'neo-colonialism' in agro-export investments. The mercantilist plans of Chinese, Saudi and other nationals of rich food deficit countries to lease vast tracts of African land for the export of staple foods back to their homelands are especially troubling for the food security of import reliant peoples (Blas 2008). The poverty impact of the very real prospect that African governments have entered a race to augment existing investor-friendly measures with other inducements to capture new investments of this type is as yet unknown.

What is much more clear is the general inability of the Tanzanian Government to guarantee that agricultural investments will have positive anti-poverty offshoots under the WTO Agreement on Trade-Related Investment Measures (TRIMs).[22] Linda Weiss (2005: 724) has contended that the measures prohibited under TRIMs are of a diminishing level of importance to the richest countries. Permissible programmes, funding and other investor supports often exceed the limited financial resources of the poorest states. As Tanzania no longer enjoys the power to impose local content or trade balancing requirements on its investors, it could not move to restrict the ability of investors on any future outgrower cotton project to procure imported pesticides. It might not even be able to mandate the sourcing of domestically-produced botanical pesticides. If the Government entered into a joint venture or public-private partnership with a foreign firm to execute the marketing and foreign sales of Tanzania's own-design cotton garment output it would also face a difficult path. Owing to the prohibition of foreign exchange balancing it would be unable to include contractual language that obligated the counterparty to ensure that its operations generated a net inflow of foreign exchange. Any Chinese investor who sourced Asian origin synthetic fabric exclusively and established an assembly plant that subsequently flooded the local market would also be immune from 'nationalist' attempts to impose quantitative restrictions on his imports or domestic sales.

On the other hand, TRIMs enables countries with the means to do so to promote the interests of their traders and foreign investors. Chinese Premier Wen Jiabao visited Tanzania in June 2006 as part of a seven-nation TRIMs-compliant trade mission. On this trip he did not simply offer President Kikwete a noteworthy package of unilateral concessions on Tanzania's exports. He made considerable efforts to advance the interest China's textile industry has in securing a greater volume of Tanzanian lint exports. Mr Wen also hoped that Chinese investors in the Tanzanian textile sector would continue to enjoy the benefits of export-processing zones and other government concessions. Beyond the trade mission, the TRIMs agreement also permits China to offer its garment exporters concessional or soft loans to augment their export capacities. Significantly, the massive public investments China has made in its biotech seed industry and in its Bt cotton project are TRIMs-compliant (Karplus & Deng 2006). Venture capital, comprehensive export readiness programmes or market intelligence services that China provides to any future spin-offs from this public research that are able to enter the Bt seed trade would be similarly acceptable. Government of China investments in the intellectual property infrastructure that are deemed necessary to ensure that these firms can patent their seed varieties and have access to domestic and global recourse to protect their patents from infringements would be similarly above-board.

A criticism John Kenneth Galbraith (1973) levelled at attempts to construct one-size fits all theories of the firm in an era of unprecedented and highly concentrated corporate power can consequently be applied to TRIMs. Regardless of a lack of malicious intentions or design, it has 'concealed the disadvantages of the weak', deprived them of development tools used elsewhere and preempted their empowerment.

Lint exports

Beyond the global oversupply conundrum, lint-forward challenges constitute a final dynamic issue area that will bear increasingly upon poverty outcomes if existing local textile capacities remain underutilized and opportunities for value addition go unrealized.[23] Value chain theorists have argued that Tanzanian and African lint suppliers in general have faced relatively less stringent product standards, fewer rules and lower transaction costs than firms that supply the continent's other traditional export commodities to international traders (Gibbon & Ponte 2005).[24] In this view, the cotton value chain is far from the quintessential buyer-driven chain defined by Gereffi and Korzeniewicz (1994). Designers, garment manufacturers, weavers, spinners and international cotton merchants have not shed processes upstream or dictated costly best practice prescriptions to the detriment of African exporters. Principal cotton merchants including Cargill, Olam and Paul Reinhart have simply sought to maximize the volumes and varieties of cotton that they procure. While the chain is hierarchical insofar as there are only a handful of major traders, spinners can cut these traders out and source directly from ginners that have significant volumes of a desired national origin on hand. The big merchants have countered the weakness of their intermediary role through establishing backwards linkages. They have pre-financed agents to deliver specific quantities, constructed ginneries, designed and implemented input distribution systems and in southern Africa, have even successfully piloted contract-based farming schemes.

The pitfalls of over-reliance on lint exports could stem from new technologies, market segmentation, entry barriers and a potential downgrading of the physical properties of lint relative to ethical appraisals of the productive relationships that generate particular fibres. Tanzanian exporters need resources that can be used to obtain the so-called high-volume instrument (HVI) technology developed by the United States Department of Agriculture (USDA) and the Institute now known as Cotton, Incorporated. The HVI enables a rapid determination of the physical properties of lint. Without it, ginners will be unable to fetch the premium associated with direct sales to foreign spinners (Larsen 2003, 2004).[25] Given that the national origin has to a certain extent fallen into disrepute, the recent polarization of the Cotlook 'A' and 'B' Index prices also does not bode well. The growing spread between these indices and the prices offered for extra-fine cottons and other highly regarded origins slated for blending also imposes a ceiling on the foreign exchange earnings available to be invested in supply chain upgrades. Market segmentation and surging consumer demand for higher-quality fabrics in emerging markets could be a recipe for industrial stagnation in Tanzania and a source of relative impoverishment over the long term. Bearing in mind Gibbon and Ponte's analysis of the entry barriers facing Africa's lint suppliers, unless Tanzanian firms and other African ginners were to form a consortium they would be unable to procure the volumes and varieties of origins required to break their position in the chain and become competitive cotton merchants. However, even this improbable development would be doomed if the continent's conventional ginners did not heed the growing body of evidence that lint quality is no longer the sole or principal determinant of value operative in the global cotton chain. Processes and production methods have assumed new importance. Conventional understandings of cotton 'quality' are less significant in the determination of 'value' for some of the newest and most high value non-traditional cottons. Conventional cotton could be heading the way of the dodo.

Views from the farm: Cotton, poverty and beyond

As field research for this case study was being carried out between March and April 2007 the Research and Analysis Working Group of the MKUKUTA Monitoring System conducted a major survey of Tanzanian opinions on poverty (URT 2007). Over 7879 people were interviewed individually and in focus groups for this work. Results were subsequently presented to officials, advocates and the research community as the *Views of the People* on poverty challenges, changes in standards of living and the state of access to economic and social services. Many rural dwellers surveyed considered the poor condition of roads to be the biggest impediment to poverty reduction. They also told researchers that the cost and availability of services such as agricultural inputs and healthcare were serious problems.[26] Well over half of the respondents claimed to have suffered from malaria during 2006, and most stated that they had not been able to purchase chemical pesticides or inorganic fertilizers. Only 55 per cent said that they had consistently consumed three meals per day. When asked if things had improved over the previous three years, more than half of the interviewees asserted that their economic situation was worse. Two-thirds argued that the cost of living was increasingly steep. Despite official stasis in the GINI coefficient, this exercise found the perception that inequality is on the rise to be pervasive in rural areas (World Bank 2007). The concentration of employment opportunities and access to goods such as cell phones and charcoal in the cities could have contributed to this grassroots take on the direction of the trend. Low levels of trust in public officials and the fact that only 22 per cent of rural people have participated in local, ward and district level planning processes could also have contributed to the view that disparities were on the rise.

These people, others whom were surveyed during the previous Participatory Poverty Assessment (PPA) and my own interviewees identified and stressed several impoverishing relationships that will be useful for stakeholders to take into consideration as they endeavour to establish poverty eradication priorities for the sub-sector. Agriculturalists are very concerned with the stress that elder care places on their resources. They would like to see the provision of services for the aged improved significantly. Women are also keen to gain control over the earnings from their labour, and many want to enjoy greater levels of physical security in their homes and communities. Several women and men told me that they were afraid to join Savings and Credit Cooperative Societies (SACCOS) owing to the fact that the size of the deposit required outstripped the cash they had on hand, the harsh terms and conditions of membership or the disproportionate control powerful farmers exerted over these societies. Based upon their responses, further investments in microcredit and finance could have high opportunity costs if accessibility issues persist or if the distribution of benefits within particular SACCOS is skewed. Another input provision problem was brought home when several of my interviewees told me that they considered me to be the first person to have ever visited their farm to discuss cotton. Finally, members of a primary society that has morphed into a buying agency for several traders discussed at length two unique marketing problems that have short-changed the society. In their words, there

> are some companies that have problems with their scales. Before we buy cotton from the farmers we send our scales for rectification and stamping and then we bring them to the village. If we measure ten tonnes here these companies say it is nine tonnes [when they weigh it at the ginnery].... Company representatives have also changed prices offered without informing their employers of these changes. When the primary society representatives go to the companies after the season to claim [payments] they are usually told that the companies are not aware of any new prices.

Each of these insights demonstrates the value of the participatory approach and the imperative for action. The hope is that attention will be drawn to the relationships people themselves have highlighted as attempts to realize the anti-poverty prescriptions detailed here or elsewhere proceed. Regardless of the ways the five dynamic factors I have discussed might limit or enable pro-poor outcomes, perspectives from the farm are indicative of the views of the true 'principals' of the reform agenda. Individuals and members of groups involved in creating and implementing the ways forward who do not reside on farms or cultivate cotton must never forget their place. They are only the agents of change. If or when the agents become the primary beneficiaries of processes that ostensibly aim to improve the human condition for the principals, poverty maintenance is assured. How well several of these agents are living up to this imperative as they push for the implementation of new principles, norms, rules and decisionmaking procedures at the global and domestic levels is the subject of the next chapter on non-governmental advocacy.

5
NGOs, Conventional Production and Poverty

The Fourth WTO Ministerial Conference was scheduled to be held only two months after the 9/11 attacks at Doha, Qatar from 9 to 13 November 2001. Given the heightened security considerations, questions were raised during the preparatory phase about the ability of non-governmental organizations (NGOs) to offer advice to participants, articulate alternatives and monitor the multilateral trade negotiations. The local restriction of civil liberties such as the freedoms of assembly, association, press and speech also did not bode well. Additionally, NGOs faced a stark public relations challenge. In the aftermath of the now Hollywoodized 'battle' on the streets that occurred during the previous Seattle Ministerial, the WTO Secretariat had stepped up its efforts to augment public knowledge of the benefits of trade liberalization to counter non-governmental voices. The Secretariat attempted to remedy what it deemed to be the widespread 'public confusion' and 'apprehension' about globalization derived from 'ill-informed comments and misinformation' available on the Internet (WTO 2000: 4; WTO 2001: 2).

At Qatar delegates extended their work into a sixth day in order to advance the view that negotiations were viable and in the interest of the global public. They sought to launch a new round that would aim to reach an agreement that would enable all member countries to reap gains. Participants were especially attentive to the concerns developing country trade negotiators had aired since the conclusion of the Uruguay Round. The Ministerial Declaration of 14 November that launched the Doha Development Agenda enshrined the latter focus. It also sent a message to non-governmental observers that WTO members had reaffirmed their faith in the legitimacy of the multilateral trading system and in the institution charged with its governance.

Many advocates, capacity-builders, researchers, service-deliverers and other concerned global citizens subsequently committed themselves to

ensuring that Doha would live up to its billing as the launch of a development round. Their efforts to coalesce around the resolution of compelling
trade and development issues were almost immediately enhanced when a
gripping tale of trade-related hardship came to light only one week after
the Ministerial Declaration was signed. That week, leaders of the national
cotton producers' organizations of Bénin, Burkina Faso and Mali had
gathered at Bobo-Dioulasso, Burkina Faso to discuss the sharp decline in
the world lint price that year and what could be done to arrest the trend.
On 21 November, Issa Ibrahim, François Traoré and Ampha Coulibay
issued an appeal to the United States and the European Union to put an
end to their cotton subsidy and support schemes. African cotton growers
believed that these systems had glutted the global market for cotton lint
and consequently reduced the world cotton price. These leaders thought
that the abolishment of subsidies elsewhere would raise incomes and
alleviate mass poverty and hunger for the 20 to 30 million Africans that
depend directly or indirectly on cotton for their livelihoods.

Northern-based groups such as Peuples Solidaires (2002) disseminated
the producers' call for 'fair' trade. A letter writing campaign commenced
early the following year to inform then European Trade Commissioner
Pascal Lamy of the plight of Africa's *coton-culteurs*. In April 2002 Oxfam
International also drew attention to cotton 'dumping' when it launched
its global *Make Trade Fair* campaign. At that time Oxfam released a publication that situated the challenges that poor countries faced in the trading system, including cotton (Watkins & Fowler 2002). Its work on the
issue specifically continued and that September, Oxfam published a report
entitled *Cultivating Poverty: The Impact of US Subsidies on Africa* (Watkins
with Sul 2002). This document was released on the same day that Brazil
filed its complaint to the WTO Dispute Settlement Body on US cotton.
Oxfam subsequently issued a press release on the topic in conjunction
with Environnement et Développement du Tiers Monde (ENDA), a Dakar-
based NGO, and the West African producers. These organizations supported the Brazilian complaint and the subsequent moves Bénin and Chad
made to become third parties to the dispute.

Later in 2002 the Government of Switzerland supported the work
of an independent, Geneva-based non-profit organization to research,
aggregate and articulate the trade reforms West African administrations
had prioritized. This organization, the IDEAS Centre, was then chaired
by former GATT Director-General Arthur Dunkel and was under the executive directorship of former Swiss trade negotiator Nicolas Imboden. The
Centre had identified the cotton emergency as a principal trade concern.
It used the Swiss funds to commence a project to advise select West

African governments on how to remedy the subsidy issue and advance their interests at the WTO. IDEAS Centre's first action was to organize a series of meetings between the International Agency for Trade Information and Cooperation, the International Centre for Trade and Sustainable Development (ICTSD), the Advisory Centre on WTO Law and Geneva and Brussels-based delegates from the sub-region. This process led to the identification of cotton-dependent countries that could support a submission in defence of their interests and to the formulation of the *Sectoral Initiative in Favour of Cotton*. Further technical assistance and several country-level interventions by IDEAS enabled participants at an Economic Community of West African States (ECOWAS) Ministerial meeting held at Accra on 24 April to endorse this Initiative. Bénin, Burkina Faso, Chad and Mali – the Cotton Four (C4) – formally submitted it to the WTO on 30 April (OECD 2006: 110).

Academic observers have lauded the ensuing engagement of ENDA, ICTSD, IDEAS Centre, Oxfam and the producers' organizations based in the Cotton Four (C4) countries to advance the Initiative and raise the global profile of the dumping issue. In the conclusion to their edited volume on the challenges facing African cotton, Leslie Gray and William Moseley noted that powerful constituencies in the North had successfully kept the issue of agricultural subsidies off the negotiating table until recent years. In their view, African governments, NGOs, producers' unions and the press 'all seem to have become more effective at influencing debates on international trade' (Moseley & Gray 2008: 280). However, these observers did not establish the latter argument empirically. They also did not stop to ask why the influence that they assumed had emerged, or to clarify why it was significant as regards efforts to eradicate the historic relationships between cotton and poverty. Gray and Moseley simply characterized the inclusion of this issue in the Doha Round as a power shift. The analysis of NGOs presented in this chapter demonstrates the paucity of their assertion. It underscores the point that a net positive impact of the new non-governmental interventionism on the historic relationships between cotton production and poverty cannot be assumed *a priori*. While African cotton producers first articulated concerns with trade, cotton and poverty, other compelling issues that impoverish Africa's *coton-culteurs* have been sidelined since Northern NGOs embraced and prioritized the trade issue. Northern efforts have not as yet fully transcended the dark conclusions that Alejandro Bendaña (2006) has articulated on the association of Northern-driven NGO awareness-raising campaigns and capacity building efforts with North-South power asymmetries. That said, ongoing engagements on cotton show that particular Northern disbursements and capacity building efforts in this area have exuded the principle of development effectiveness. Despite the litany of failings articulated below, Northern NGOs have in certain contexts mobilized Southern capacity and fuelled South-South cooperation. Consequently, the snapshot of Northern NGO engagement on cotton offered in this chapter is a highly nuanced one.

Taking Scholte's (2002) analytical framework as a point of departure, the scorecard of non-governmental influence is actually much more balanced – and problematic – than adherents to the unidirectional (blanket positive) view imply.[1] It is clear that the new voice that has been given to 'reformist' perspectives has been effective insofar as decisionmakers are now aware of the need to alleviate the oversupply crisis. They are more knowledgeable about the poverty challenges associated with commodity production. Celebrity campaigns and the global media attention orchestrated by non-governmental activists have also been effectual. They are amongst the sources of a new consumer curiosity in the relationships of production and exchange that underlie everyday consumables that have been derived from African produce. Even my own interest in understanding and explaining the cotton conundrum can be viewed as derivative of these awareness-raising efforts. NGOs can take some credit for the fact that progress on the cotton file is now considered a *sine qua non* of any Doha final agreement, should one ever materialize. As a result of non-governmental actions farmers and the C4 governments have also started to reap gains. New education programmes (including the new University of Cotton in Bobo-Dioulasso), aid disbursements and cotton-specific projects have materialized since Blaise Compaoré, the President of Burkina Faso, presented the Sectoral Initiative to the WTO Trade Negotiations Committee.[2]

On the other hand, it cannot be said with certainty that NGOs have had a significant effect on trade governance. While the United States lost its final appeal in the Upland cotton dispute in June 2008 it is by no means clear that current or future US administrations will move to fully eliminate the aspects of the US cotton system that the WTO Appellate Body found to contravene WTO law. It is also possible that the pursuit of improved market access, subsidy removal and dedicated development assistance have obscured other trade-related problems. Northern NGOs have not been as vocal about price volatility, secular price declines, or the challenge of improving opportunities to add value to cotton locally. Factors not directly related to trade that bear upon poverty reduction have also been out of sight. For example, many non-governmentals have continued to accept and propagate the goal of liberalization while not devoting similar resources to reforming the global exchange rate regime. Despite a recent Euro-dollar trend that reduced the incomes of the ostensible 'beneficiaries' of NGO cotton activism in the Communauté Financière d'Afrique (CFA), no transnational campaign has been launched to raise awareness of the impacts that fluctuating currency valuations can have on poor countries. At best, these organizations have been co-opted into a political project that enhances the legitimacy of free trade liberalism while continuing to draw attention to broader poverty challenges. At worst, they have acted explicitly as agents of the institution charged with governing the multilateral trading system. They have secured the conformity of African officials to the status quo norms, rules and decision-making procedures of trade negotiations, and sought to professionalize and depoliticize African approaches to trade and development debates.

Overall, since farmers first articulated the cotton problem a transnational activist and policy elite largely based in the North has dominated collaborations to formulate strategies and execute tactical actions to better the lot of *coton-culteurs*. As yet, no mechanisms have been developed for cotton growers to hold participants in this transnational network accountable for their actions. Moreover, prominent members of the various non-governmental coalitions operative in the network have managed what David Hulme (2008: 38) refers to as the trade-off between their own 'institutional imperatives' and 'development imperatives' largely with an eye toward the former. In Alejandro Bendaña's (2006) words, these organizations have chosen to contest policy rather than power, and to undertake 'capacity building' instead of a more thoroughgoing 'capacity mobilization'. Perhaps most damningly, the distribution of the flows of investments and services that the network has stimulated remains highly skewed. As such, just how the millions of cotton-dependent producers in non-C4 countries will benefit from networked activism for policy reform remains unclear.

To develop the above argument the next section discusses how the network, coalition and social movement concepts can be applied to this issue area. It offers a brief analysis of the norms underlying the political engagements of the two key transnational coalitions that have worked on the topic. The following section details how Oxfam's choice to hone in on liberalization was internally controversial from the outset. This orientation has compromised the development effectiveness of its cotton campaign even as its research outputs and work on regional trade integration in West Africa have highlighted other factors of impoverishment. Next, the IDEAS Centre's top-down capacity building projects are explained and their impacts on the potential for poverty eradication evaluated. A final section charts the anti-poverty effect that found-ations, civil society advocates, non-governmental service-deliverers and community-based organizations are having beyond the ambit of the cotton network. To do so a brief case study analysis of Tanzania is pre-sented. This section underscores the marginalization of the country's Western Cotton Growing Area *vis-à-vis* other regions that are more geo-graphically proximate to the hubs of the non-governmental aid delivery system. The conclusion resonates with Sangeeta Kamat's (2004) finding that NGOs cannot automatically be labelled 'innocent'. Mirroring David Hulme's analysis of global NGO impacts, to date the cotton network has not facilitated Southern empowerment or built a constituency for real change in global consumption and production patterns. Cotton farmers have simply not been able to hold the network's key decisionmakers to account.

Networks, coalitions and the transnational norm for cotton

In the wake of the Bobo-Dioulasso appeal a transnational advocacy net-work emerged. Non-governmental advocates from the South and the North, representatives of producers' unions, trade experts, academic econ-omists, professional researchers, international civil servants, donors, dev-eloping country officials, industry association staffers and others became increasingly linked in a transborder exchange of information (Khagram et al. 2002).[3] As Margaret Keck and Kathryn Sikkink (1998) might have predicted, the Internet enabled these elites to come together across bound-aries and engage in an extended discourse. The research contributions and informal dialogue that flowed through this network aimed to improve knowledge of the situation on the ground. Participants discussed what was being done and what could be done to realize a fairer cotton trade (Hazard 2001, 2002). Most contributors sought to build this collective belief into a 'standard' of appropriate behaviour held by a critical mass of states or an 'international norm' (Finnemore & Sikkink 1998). While the nature of the network was such that participants could benefit from recip-rocal interactions seemingly as equals, hierarchies along the lines of those Sikkink (2002) had identified in other networks lurked in the background. These came to the fore only after certain groups took steps to establish a division of labour and coordinate their tactics to advance the new norm. In so doing these entities fuelled the emergence of transnational advocacy coalitions that aimed to publicly influence social change.

Oxfam International spearheaded the first such coalition. From the outset its *Make Trade Fair* campaign necessitated the transboundary coordination of research and advocacy efforts. This work entailed the identification of complementary trade programmes within the various national Oxfam organizations and the alignment of these with the new global action. For example, after the launch of the campaign, Oxfam Great Britain's activities to inform, organize and generate the heightened political participation of West African farmers became a vehicle to raise official awareness of the subsidy issue. Working with ENDA and producers whose national organizations fell under the umbrella of the Réseau des Organizations Paysannes and de Producteurs de l'Afrique de l'Ouest (ROPPA), Oxfam GB subsequently raised the cotton crisis with members of the secretariat of the Union Économique et Monétaire Ouest-Africaine (UEMOA) and the Conference of Ministers of Agriculture of West and Central Africa. This NGO also supported the linking up of national social movements on the topic.[4]

Nicholas Imboden (2004) later summed up the division between this coalition and the operations of his own organization and its Geneva-based collaborators and 'partners' in a presentation entitled *Société Civile et OMC (WTO)*. He delivered this message at a roundtable on cotton hosted by ICTSD. Imboden characterized Oxfam and other NGOs with a similar orientation as groups that had 'sounded the alarm' and engaged in successful momentum generating 'sensibilization'. In his view these actions had laid the groundwork for the IDEAS Centre to enter the picture and take the issue to the next level. IDEAS was portrayed as an apex institution charged with counselling African leaders and assisting the construction of an international action plan. He depicted the Centre as a great enabler. In this light, the Centre's provision of technical assistance made it possible for cotton-dependent countries to make their case known at the forum where trade decisions were actually made. Oxfam's parallel and implicitly subordinate role was to simply keep the political and media pressure up while the 'technical' work was ongoing. By declining to label Oxfam a 'research organization' and by classifying only the Overseas Development Institute as a non-governmental of this type, Imboden reinforced the notion that Oxfam's contributions, if not secondary, were at least delimited to trade 'politics'. He considered the latter discursive realm to be distinct from trade diplomacy and the specialist language deployed during negotiations. Imboden's overall strategic message seemed to be that a non-governmental strategy harmonized with the prevailing norms and procedures of trade governance and led by a small, well-connected group would be most likely to yield results. As such, his expectations contrasted with those held by at least one expert.[5]

The coalition of elites that came together to back the IDEAS Centre's purposive intervention in favour of trade liberalization and to take delivery of its advice was nonetheless a transnational one (Scholte 2007). This characteristic has been evident insofar as official European donors have financially backed the efforts of a non-governmental entity to augment the capacity of several African governments to engage in trade negotiations. The coalition has demonstrated a level of transborder cohesion, generated and disseminated information, and facilitated the provision of education. IDEAS staffers have predictably described the North-South flows of finance, technical assistance and trade-related capacity building that they have overseen as 'necessary' (Imboden 2004). They could also reasonably claim to have advanced the agenda articulated by the Panel of Eminent Persons on United Nations-Civil Society Relations within the trading system. This panel had been struck to redress the 'imbalance between the voices of Northern and Southern actors' in policy forums (United Nations 2004: 29).

Nevertheless, the transplanetary quality of this alliance must not be overstated (Löfgren & Thörn 2007). Asymmetries have persisted within this transnational coalition that can be framed in North-South terms. Owing to the evident lop-sidedness the norm underlying the coalition's work was narrowly conceived at the outset and imposed in a top-down fashion. As a consequence, the benefits set to flow from its work to initiate the process Sikkink has labelled 'norm shift' have been circumscribed.

In fact, the transnational norm adopted and brokered by IDEAS and by Oxfam was one that many heterodox development economists would be adverse to. If it ever achieved the status of a shared expectation of appropriate behaviour for rich and poor countries alike in the international system, trade and development debates could persist indefinitely. The norm espoused by both coalitions cannot simply be summarized as an appeal to *make the cotton trade fair for Africa*. In more precise terms, the coalitions sought to advance the belief that *liberalization of the cotton trade is necessary and just for African poverty reduction*. This norm contains a statement of cause and effect, and also an appraisal of right and wrong. Consequently, it bedevils the distinction between causal and principled ideas typically upheld by experts who analyse prospective norms (Khagram et al. 2002: 11). The statement of 'oughtness' – a standard component of international collective beliefs – was in this instance coupled with a proposition that could be considered over generalized and ill-defined. It seemingly endorses Africa's immiserating dependence upon commodity exports. Oxfam itself confirmed that the appraisal at the heart of the norm was controversial when an internal conflict broke out subsequent to the Bobo-Dioulasso appeal. Several of its country-level staff were wary of a sole focus on liberalization or dumping, and later struggled with the level of emphasis to place upon it during the global campaign.

After the norm was embraced and prioritized Oxfam's advocacy work and the ensuing media saturation enabled it to reach a threshold or 'tipping point' in the transnational network (Finnemore & Sikkink 1998). Beyond the US, officials and non-governmental actors alike internalized the norm. It became further entrenched as IDEAS pursued its agenda. The Centre's efforts solidified the terms of the debate and showed the power of the norm to channel and regularize behaviour on the cotton issue. As such, through seeking to foster trade policy change or augment the capability of others to do so, both coalitions had not simply adopted a 'reformist' orientation, but also a highly 'conformist' one (Scholte with Schnabel 2002). The following analysis of both groupings demonstrates that this norm and associated political developments have to a certain extent been detrimental to African growers and their movement. As the centrepiece of calls to action it has precluded what Bendaña (2006) has termed 'heightened political awareness' or what Paulo Friere might have referred to as the 'conscientization' of people about this problem.

Oxfam cottons on

In 2001 Oxfam International set out to implement a five-year strategic plan. The principal ambition of this plan, known as *Towards Global Equity*, was to achieve considerable, sustained and multiple impacts on the lives of very large numbers of people. To realize this outcome Oxfam set out to promote changes in policies, global public opinion and in learning. In particular, its strategy mandated the devotion of the maximum levels of attention, energy and resources possible toward establishing 'fair rules' for the global economy. The organization gave the 'highest priority' to increasing the access poor people had to markets, empowering the poor in market relationships and reforming global trade. This action plan took concrete form in the Make Trade Fair campaign's emphases on improving access to medicines, eliminating dumping, enhancing fairness in trade and promoting the recognition and enforcement of labour standards and rights. Regarding what was internally referred to as the cotton dumping campaign the objectives were to bring about the end of Northern subsidy regimes and to ensure that a transitional compensatory mechanism was established. Oxfam also wanted trade negotiators to treat cotton separately and hoped that cotton-dependent countries would take delivery of more external resources.

An external evaluation team was charged with reviewing the implementation of the dumping campaign in 2006. It concluded that most of Oxfam's objectives in this area had not been achieved. The subsidies were still in place and compensation had not flowed. My research indicates that this lack of traction was rooted in the fateful decision Oxfam made after a series of external interventions and a heated internal debate in 2002 to throw its support behind the then embryonic trade negotiations strategy. Prior to this decision, Oxfam had worked with ENDA to organize multistakeholder meetings on the topic. These were held at Lomé and Abidjan, and had aimed to facilitate producer participation in defining the sub-region's response to the cotton crisis. Ministers of agriculture from West and Central Africa had been apprehensive about this new cross-border, NGO-driven organization. Even so, they emerged from their ministerial conference in 2002 ready to take up the issue at the WTO, and dispatched a letter to then WTO Director-General Mike Moore. His reply informed the ministers that they had no standing at the WTO and outlined the proper procedure necessary for member countries to raise their concerns.

That June, Eric Hazard, then a researcher for ENDA, and the Geneva-based head of the Make Trade Fair campaign penned a note that detailed the alternatives African governments had moving forward. Amongst other options, this memo noted that they could become third parties to the looming Brazilian complaint on US cotton at the WTO. However, the list did not include the idea that they should pursue a remedy through trade negotiations. According to my interviewees this alternative simply had not occurred to the authors. It was only several months later that this possible direction came to be discussed as a direct result of Nicholas Imboden's efforts to promote the political argument that negotiations were the right course of action. High-level members of the campaign and certain Oxfam affiliates were not keen on Oxfam publicly supporting this route. The belief that African governments, with assistance, could advance their own agenda through negotiations and build alliances in a way that was 'too good to miss' was not broadly shared in-house. Several insiders have contended that a minority believed negotiations to be a pro-market, liberal orientation that did not accurately reflect Oxfam's wider agenda. While many of these voices took issue with the way the *Sectoral Initiative* was subsequently formulated and criticized its emphasis on subsidies, Oxfam 'went along with it'. Once the decision was made to run with a negotiation strategy that had 'swooped down from the top', the regional consultations on the ways producers preferred to address cotton became imbued with the new imperative. Thereafter, Oxfam's cotton campaign was brought into *de facto* alignment with the means that the IDEAS Centre had proffered.

After pursuing this course and deploying tactics such as celebrity, citizen and event-based campaigning, Oxfam conducted an internal evaluation of the progress it had made toward implementing the strategic plan. This survey identified a sense within the organization that the emphasis on liberalization had been imposed from above. Nonetheless, its authors concluded that many staffers considered cotton to have been the most successful Make Trade Fair effort to that date. In their appraisal, cotton had been launched from an obscure concern into a symbol of the inequities of the world agricultural trade and a litmus test for the realization of global justice. While the report conceded that it was 'premature' to expect direct impacts on peoples' livelihoods, the authors celebrated what they deemed to be the campaign's significant impact on attitudes and beliefs. On the whole the internal evaluation characterized Oxfam's approach as a 'demonstrable' success.

The independent external evaluators subsequently noted that the latter assertion 'seemed rather far-fetched'. They also argued that it was somewhat disingenuous to assert that the campaign had shifted public views, as this claim could not be verified with any degree of certainty. The evaluators lauded instead the campaign's successful efforts to inform people about the links between trade policies, practices and livelihoods. They also encouraged Oxfam to ask whether or not its advocacy work was helping to build the capacity that would be necessary for a sustained effort to change trade policies and practices. Additionally, they wondered if the campaign had generated any unforeseen trade-offs.

The slightly ironic view that the campaign had been 'successful' was not only upheld by employees with full knowledge of its unmet objectives, but also by external stakeholders who were approached to give their opinions on the matter. These individuals were positive about what they perceived to be the heightened public awareness of dumping and its impacts. Others asserted that Oxfam had seemingly helped to shift the debate on trade policy change toward development issues. That being said, they also felt that Oxfam's campaign model 'oversimplified' a complex topic and entailed a significant level of risk. The organization had staked its reputational capital on cotton, and interviewees worried that it had not adequately managed expectations. Nearly all respondents opined that there had been no discernible effect on cotton poverty in West

Africa or that it was 'too early to tell'. Interviewees rated the effective‐ ness of Oxfam's interactions with delegates to the WTO and the US administration extremely low. It is possible that the slow or deficient uptake of the subsidy message by Northern officials and the persistence of *Realpolitik* were taken by many to absolve Oxfam for failing to meet its targets. If the 'messenger' had indeed been disinterested, surely it would have been unwarranted for these people to shoot the messenger for the failure of the recipients to heed the message. However, the mes‐ senger in this instance was responsible for the contents of the com‐ muniqué. As such, no matter how it is spun, this agency is responsible for the diversion of development finance into a strategy that is failing or has failed. Power politics and the big powers alone cannot shoulder all the blame.

That being said, it is important not to throw the baby entirely out with the bathwater. To its credit Oxfam's own research (Baden 2004), studies that it has supported and high-profile events that it has backed have drawn attention to aspects of the cotton problem beyond liberal‐ ization and the need for development assistance. For one, Oxfam backed Eric Hazard's (2005) work. His edited volume shed light on the price issue and the development aspect, and also detailed country-level and sub‐ regional governance and organizational challenges not covered by the submission. Other researchers have subsequently taken up these issues. ENDA Diapol's new research on cotton focuses on the poverty impact of the removal of intra-regional trade barriers and the ways to increase pro‐ ducer participation in the design and implementation of policy. Hazard's collection was initially presented on 12 December 2005 at the Hong Kong Ministerial during an event known as 'Cotton Day'. ENDA, ICTSD, Oxfam and the Permanent Mission of Bénin in Geneva had supported this event to facilitate the exchange of dialogue on the Initiative and on the cotton dossier more generally.

Still, support for comprehensive research and this one-off event are the only instances where the road not taken has percolated to the surface. The campaign has otherwise been dominated by high profile and costly attempts to maintain dialogue and attain influence, or to spread the word about the effects of dumping in West Africa.[6]

Free trade liberalism, WTO legitimacy and the IDEAS Centre

The IDEAS Centre describes itself as an organization that offers 'practical, results-oriented' advisory services. It claims to execute 'projects aimed at strengthening the capacities' of developing country governments to 'shape both their domestic economic policies as well as the international policies that affect them'. The Centre has encouraged its clients to use their WTO membership in a way that promotes 'sustainable human development'. It has worked to deepen the understandings that these governments have of the linkages between development challenges, international trade and WTO rules. The core belief that underlies its efforts to improve coherence between trade and development policies and to scale-up the trade negotiations capabilities of poor countries is that 'free market forces' and the 'rules of a liberal world trading system' can be harnessed or directed toward the achievement of 'poverty alleviation'. This conviction has been especially evident in the Centre's focus on the cotton problem.

From January 2003, when Switzerland backed the Centre's efforts to provide technical assistance to cotton-dependent countries, it has explicitly sought to delimit discourse on cotton and development to the trade issue.[7] As the cotton project has progressed through four distinct stages it has functioned as a mechanism to promote the conformity of poor governments to the rules of the multilateral trading system. The IDEAS Centre has driven the provision of assistance and trade-related capacity building over the years and brought about North-South transfers of knowledge, skills and resources. These actions and flows have created a new cadre of trade experts and expanded the representation of targeted governments in Geneva. Nevertheless, the Centre's political strategy has thus far not enabled these countries to bring about the changes that they desire. IDEAS has failed to enhance their competencies to analyse and act upon all but one aspect of the historic relationships between cotton and poverty.

It is more accurate to characterize the assistance that the IDEAS Centre supplied during the first and crucial phase of the project as political rather than technical. Nicholas Imboden engaged in a concerted and expensive lobbying effort to convince heads of state and government of the need to pursue a trade negotiations strategy. His interventions did not simply aim to augment the problem-solving capability of the governments concerned. He came to West Africa with a particular view on how these countries could increase their effectiveness and forcefully brokered this perspective in the hope that a group of states would adopt his preferred means to stick their heads above the parapet in Geneva. These efforts brought the governments of four cotton-dependent countries into alignment with a political solution that had been envisaged in the North. Far from simply refining the Initiative's technical aspects, or as the Centre put it, analysing the 'trade interests of the region' through innocuous 'desk research' and 'visits to the field', Imboden conceived this approach to helping countries 'defend their interests'. He personally facilitated its uptake by the four countries. IDEAS subsequently contended that one of the successes of the first phase was that the C4 was able to credibly and forcefully defend 'their' Initiative at the WTO and in so doing attain ownership over it. While technically correct this appraisal obscured further political manoeuvering that had taken place behind the scenes after the Initiative was agreed.[8]

IDEAS continued to dominate the formulation of actions taken in support of the submission after a July 2003 meeting of bilateral donors launched what came to be known as the 'cotton emergency project'. Switzerland, Denmark, France, the Netherlands, Sweden and the United Kingdom disbursed a total of 608,000 Swiss francs (CHF) to support the second phase. In the lead up to the Cancún Ministerial this project supplemented the advice and negotiation guidance IDEAS was able to provide through the still ongoing first phase. It enabled additional research and made a high-level pre-ministerial event possible. These funds also covered the C4's additional travel expenses, enabled Bénin, Chad and Mali to temporarily station one additional diplomatic staff member each in Geneva, and backed a communications strategy. The latter burst into the public view on 11 July when an opinion piece attributed to Presidents Touré and Compaoré appeared in *The New York Times*. Unfortunately, its provocative title – 'Your cotton subsidies are strangling us' – and its contents did not generate the hoped-for controversy. Instead, this consciousness-raising move fumbled when it came to light over the ensuing days that the IDEAS Centre had likely pursued the publication of the piece without full knowledge that President Touré had been informed of the effort. The Centre's subsequent communications with the *Times* on the matter reveal that it relied upon a green light from Mali's diplomatic staff in Geneva. Diplomats there claimed they had been in contact with personnel at the Ministry of Commerce and Industry or the Ministry of Foreign Affairs.[9]

Weeks later IDEAS furnished the then President of Bénin, Mathieu Kérékou, with speaking notes prior to his meeting with the former US President that seemingly contradicted its public relations approach. Kérékou's talking points attempted to correct the impression that the Initiative was against the US. They played up the view that it was more accurate to conceive the Initiative as an attempt to phase-out all cotton subsidy systems maintained by countries in the North and the South.[10] Micro-management and possible intrusion into sovereign affairs continued in September at Cancún. There in an unprecedented move that several other delegations opposed, Imboden worked side-by-side with African delegates to advance 'their' agenda. After the Cancún walkout several African non-governmental organizations active on the cotton file began to think that the Centre had overstepped. Several argued that its zealous quest to stimulate the pursuit of a timetable for liberalization and compensatory flows at the global level might be a 'never-ending battle' that did not fully square with producer interests (Hazard 2005).[11]

From April 2004 the emergency phase was transformed into a comprehensive, two-year technical assistance and capacity building scheme known as the Multi-donor-C4-IDEAS project. Backed with 2.65 million CHF provided by the original group plus Germany, the IDEAS Centre set out to supply analysis and technical advice to the C4's Geneva missions. This phase also aimed to help delegates identify, strengthen and articulate their positions in negotiations, and to reinforce flows of information and personnel between Geneva and the capitals. The Centre organized regular meetings, made efforts to coordinate a unified C4 position and engaged in a concerted communications strategy to influence reform proposals. An electronic newsletter was published biweekly to advance the latter cause.

Over half of the new funds were devoted to the project's well-intentioned capacity building component. Under this aspect a technical 'antenna' or training base was established in Geneva. At this hub 'cotton collaborators' or interns from the targeted governments could learn about the WTO system and the cotton trade problem, and follow and author reports on the negotiations for several months at a time. Overall, the Centre portrayed the project as the crucial link between Geneva and the capitals on cotton. In its view, the June 2004 cotton partnership between the EU and the African, Caribbean and Pacific (ACP) countries was a complementary if lower-level initiative. Similarly, IDEAS considered the interactions of governments with bilateral and multilateral donors on the topic to play supporting roles.

From the outset the third phase maintained and exuded the top-down, trade-centricity that the Centre's earlier work had radiated. At a meeting of the pilot group held on 16 September 2004 several beneficiaries questioned why the funds and activities were concentrated in Geneva and expressed their desire for a more regional orientation. Donors challenged this perspective and essentially argued that the Geneva focus was 'necessary' to resolve the problem that their 'partners' were experiencing. In offering this rejoinder the donors did not draw attention to a crucial 'fact'. They had disbursed funds to an NGO that offered advice and support explicitly guided by the 'objective of making the multilateral trading system more conducive' for African countries and building their 'confidence' in the system to 'assure the sustainability of the WTO'.[12]

This purpose was so central during the third phase that IDEAS staff downplayed the importance of cotton-specific aid projects and even characterized these as 'counterproductive' to negotiations. For example, after the EU-ACP partnership was established, in a note dated 17 June, one analyst argued that 'assistance will never solve the problems' at the root of 'the cotton issue *as described in the African submission*' (emphasis added). Rather, the problems stemmed from 'a lack of adjustment acceptance' in particular countries that no longer enjoyed a comparative advantage in cotton and were 'not related to a lack of resources or assets' on the African side. Building upon these assertions the analyst concluded that development institutions should 'play only a subordinate role in the debate on the international cotton trade'.

Such single-mindedness spilled over into the Centre's other activities, including its efforts to educate cotton collaborators and the information disseminated in its 'Cotton Update' newsletters. On the former, 16 interns were able to learn a great deal about trade governance and return home to contribute on trade and (sometimes) cotton-related topics.[13] However, they were potentially indoctrinated with a trade-centric analysis that restricted the development effectiveness of the capacity building exercise itself. It is possible that the antenna made it difficult for interns to develop the broader understandings necessary to produce comprehensive analyses of the cotton-poverty nexus. Regarding the newsletters, respondents to a 2007 survey of stakeholder views praised the content and accessibility of the updates, but also noted that attention to subsidies had detracted from other important foci.

The IDEAS Centre pushed for a one-year project extension after trade negotiations stagnated in 2006. Donors eventually granted this request and disbursed a further 900,000 CHF even as recipients raised questions about the overall direction of the exercise. The consensus underlying the fourth phase was that despite the project's faults, the protracted state of the round made it 'useful and essential' to keep it open. Donors simply did not want the C4 to demobilize, and they hoped that its members had access to support if negotiations picked up. Principal components of this phase included the training of eight more interns and the completion of an independent evaluation. Objectives remained largely unchanged, save for the inclusion of an unbudgeted commitment to extend capacity building to other West African cotton-producing countries. This shift came at the insistence of the C4. They had found it politically difficult to justify the project to other states that were similarly dependent upon lint export revenues. Nevertheless, donors suggested that the inclusion of other countries would raise financial costs and unnecessarily complicate efforts to complete the 'transaction' C4 members were trying to effect.

Before the fourth phase moved forward recipients had also wanted to see more attention devoted to the development aspect. Donors disabused them of the notion that a focus on negotiations and evaluation was of minor importance and the project proceeded as before. The Centre continued to view the provision of aid as an inferior means to resolve the crisis and riled against the use of development assistance to prop up the 'questionable quality' of state and parastatal institutions. The implicit claim here was that legitimate action on the cotton problem had to centre on the trade aspect. As such, IDEAS set itself up for failure as regards development effectiveness. In David Hulme's (2008: 38) view, the degree to which a given strategy compromises the logic through which legitimacy is claimed provides a useful test of whether organizational self-interest has subordinated development mission. Given the above analysis, IDEAS seems to be a textbook case of what not to do to advance the interests of the poor. One of the personalities interviewed by the external evaluation team alluded to this conclusion when they offered a devastating summary of the project outcomes: 'ce sont les autres qui ont gagné – les ONG, le transport aérien etc. – plutôt que les paysans'. Simply put, the farmer-advocates have benefited more than the farmers.

Beyond the cotton coalitions: Lessons from Tanzania

Cotton zones across Sub-Saharan Africa are typically remote from political and commercial capitals, and it is possible that this relatively unfavourable economic geography has impeded flows of non-governmental resources to cotton producers. My data suggests that that a new core-periphery relationship has emerged in Tanzania as these actors have assumed more prominent roles in the provision of development finance and services.[14] Given the extent of the geographic concentration of non-governmental interventions my interviewees identified, and the instances of underperformance they highlighted, more research on this phenomenon is required at the country-level and elsewhere. It is important to ascertain in concrete terms the ways that NGO investment and operations clusters have disadvantaged cotton growers, or have worked to their benefit. That being said the stories of exclusion presented below provide a cautionary tale on the limitations of the sector. The work of individuals within the NGO community to overcome the multiple dimensions of poverty has had several notable impacts, but as yet, has not made a significant dent in the factors that have impoverished farmers and families in the Western Cotton Growing Area (WCGA).

National and international NGOs and a donor-backed foundation that is now the largest grant-making mechanism in the country have simply been unable to scale-up their efforts in the Lake Zone to the extent evident in Dar es Salaam or Arusha. Envirocare Tanzania, for example, has done much to raise awareness about the dangers of pesticide exposure and the need to implement and monitor a regulatory framework for genetically modified organisms. However, their education efforts on the former have been largely delimited to coffee, tobacco and other crops grown in the north and east of the country due to financial constraints. Similarly, ActionAid has devoted considerable attention to HIV/AIDS education, facilitating the attendance of young women at secondary schools and mobilizing coffee growers and others to form and maintain marketing cooperatives. While the latter focus has helped farmers in isolated areas of Kigoma, attempts ActionAid has made to organize agriculturalists have not focused on cotton.[15]

For its part, the multi-donor funded Foundation for Civil Society has disbursed grants to community-based organizations to promote informed policymaking, governance improvements and accountability at the local level. According to Executive Director John Ulanga, the Foundation (Foundation for Civil Society 2006) has pursued this course to buttress growth and the reduction of income poverty. It also aims to improve quality of life and social wellbeing, respectively the first and second official pillars of the National Strategy for Growth and Reduction of Poverty (MKUKUTA). Through empowering people themselves to address embezzlement, remedy the diversion of resources to unproductive uses and end the disjuncture between the high growth rate evident down to 2008 and the dearth of economic opportunities in the countryside, Ulanga claims that the Foundation is giving 'voice to the voiceless'. Even so, he laments the low-level of disbursements the Foundation has been able to arrange for residents of the cotton zone. The capacity of activist community organizations remains relatively deficient in the WCGA.[16] More broadly, official donors, foundations and NGOs have not been supportive of the nascent cotton producers' organization. This neglect was evident in 2007 when farmer advocacy organizations coalesced to push for a new national fertilizer policy. Cotton growers were conspicuously under-represented.

Regrettably, at least one targeted service-delivery effort in the WCGA has been found wanting. Most interventions there also retain a short-term particularistic orientation, and cotton farmers have not reaped gains from innovative research-based anti-poverty strategies that have been implemented near the core. On the first point, several interviewees praised a clothing and mattress distribution drive that Plan International organized in Shinyanga region to assist vulnerable children. However, a subsequent evaluation of Plan's work suggested that it had not been 'strongly' effective. Children had not benefited from Plan's choice to augment their assets exclusively. Plan had not provided similar levels of assistance to other members of their households, or to their kin or communities. In general, relatives made off with the goods immediately after they were distributed and this diversion left the targets no better off than before.

Regarding NGO engagements more generally, most cotton zone residents that I spoke with highlighted the fact that these groups seemed to be primarily engaged in HIV/AIDS education and not all portrayed this involvement in a positive light. The consensus was that the Elizabeth Glaser Pediatric AIDS Foundation, the African Medical and Research Foundation and NGO-like interventions by the Tanzania-Netherlands Project to Support AIDS Control and the Tanzania Social Action Fund had helped cotton-growing communities to comprehend the dangers of HIV and to better understand prevention methods. In contrast, the most common complaint about the perceptible concentration on awareness and prevention was that it diverted the allocation of resources away from investments in dispensaries and clinics.[17] One particularly perceptive respondent wondered to what extent this focus came at the cost of the domestic development of related value-added industries such as condom factories or test kit manufacturing facilities. How this apparent single-mindedness has detracted from a drive to remedy the relative underdevelopment of microfinance institutions in the WCGA, the inaccessibility of agricultural credit guarantees there or the dearth of small and medium enterprise capacity building projects – if at all – is at present unknown.[18] The barriers that cotton farmers themselves would face if they were to ask this question or pursue similar lines of inquiry are less ambiguous. With one notable exception detailed in the following chapter, they have not been able to benefit from an effort to extend projects to the WCGA that have been piloted in the east or in other countries to enable rural people to identify their own problems and act upon them. Research on Poverty Alleviation (REPOA), a Dar es Salaam-based think-tank, has incubated one such mechanism near the coast. Their innovative approach to participatory action research has not been taken up by other organizations with the resources to apply it in the remote west.

Beyond HIV/AIDS education and other service delivery successes, the really good news on NGOs coming out of the cotton zone stems from activities undertaken to mitigate adverse events, provide disaster relief and support drought recovery efforts. One particularly serious incident occurred during the 2006–07 cotton-marketing season when no buyers arrived to purchase the crop that farmers in Tabora region had managed to grow. Several of these producers had previously been targeted by Millennium Promise's flagship Millennium Village initiative. According to one prominent cotton buyer, workers on the cluster of six such villages in Mbola travelled to Mwanza to inform the private sector and the Cotton Board about the plight of poor people who could not market their crop. That year NGOs also worked closely with each other and with the World Food Programme and regional, district and ward officials in Shinyanga to determine which areas had been hit hardest by the drought and to provide relief.

Several cotton producers told me that a school lunch provision programme Oxfam Great Britain had previously introduced and the organization's efforts early in the disaster to bring maize to the worst affected areas had been especially helpful. In Meatu, historically the most cotton-dependent district, Oxfam's emergency assistance has evolved into a comprehensive recovery programme that aims to diversify and build the assets of individuals and families that community members themselves identified as especially disadvantaged.[19] Nearly 1000 farmers have each received a basket of assets and cash relief worth 233,000 shillings, a value that considerably exceeds a typical yearly income. Beneficiaries took delivery of five goats, sorghum seeds and 108,000 shillings after making an oral agreement not to sell their animals. The hope underlying this project was that goats would function as a store of value and that this new asset class would help to improve the benefits of cattle ownership – the traditional store of value amongst the Sukuma people – during droughts.[20] In 2007 one goat recipient claimed that he no longer felt 'poor' on a day that he had brought his cattle to an anthrax inoculation drive also sponsored by Oxfam. Taken together, these interventions seem to have uplifted a small number of cotton producers in a district where linkages between poverty and cotton production had previously been entrenched.

Concluding statement

In spite of the fact that a seriously deficient transnational norm has been embraced to further the alleviation of cotton poverty, and the reality that efforts to enshrine this norm have been plagued by a lack of pan-Africanism, downward accountability and producer voice, there are grounds to believe that non-governmental actors are starting to advance a broader poverty eradication agenda that addresses aspects of the factors and necessary interventions detailed at length in Chapter 4. In 2006, Camilla Toulmin devoted her Rachel Carson Memorial Lecture to the topic of cotton, family farms and livelihoods (PAN UK 2006). The following year the International Cotton Advisory Committee (ICAC) convened an expert panel on the social, environmental and economic performance of cotton (ICAC 2007). Also in 2007, the Association des Producteurs de Coton Africains (APROCA) struck a partnership with a French management school and foundation to create the University of Cotton at Bobo-Dioulasso, an institution that aims to improve the management of pro-ducers' organizations and cotton companies from across the West African sub-region (APROCA/HEC/FARM 2007). More widely, the Global Call to Action Against Poverty, celebrity campaigning, and the finance, food and fuel price crises kept poverty on the global policy agenda and media radar down to 2010.

All the same, it is still possible that broader global non-governmental campaigns could compound the significant risks and costs associated with cotton-specific NGO involvement with African cotton. For example, the Alliance for a Green Revolution in Africa is a powerful transnational coalition that hopes to apply the 'lessons' of the agricultural 'revolution' to Africa (Rockefeller Foundation 2006). Venturesome philanthropy, astute agricultural research, the aggressive recruitment and training of scientists and farmers, and determined agricultural and water policies have been called for. At the outset, the notion that previous agricultural 'revolutions' were possibly 'evolutions' that built upon centuries of agricultural know-ledge where they occurred, and that these experiences might be inapplic-able in present-day Africa, seemed to be somewhat lost on the new revolutionaries. Without demand-driveness and local ownership, the threat that this Northern-based NGO intervention and others could enable rela-tively deficient, painfully slow progress or even regression on poverty reduction imperatives remains with us.

6
CSR and the Cotton-Poverty Relationship

This chapter evaluates the progress and pitfalls of several distinctive types of corporate social responsibility (CSR) *vis-à-vis* the cotton-poverty relationship, and also discusses country-level factors that can impede the uptake of CSR or its efficacy. It hones in on a global norm-building effort known as the Better Cotton Initiative (BCI) and on the work that has been done to establish a Cotton Made in Africa (CMIA) product label. A case study of Tanzania's organic cotton movement is presented, and an account of the ways that the conventional cotton-buying scene in Tanzania is consequential for attempts to introduce CSR and make it viable is elaborated. The chapter contends that 'hardcore' approaches to responsibility involving third-party certification have a greater poverty-reducing potential than lighter-touch alternatives. However, the more stringent CSR variants are by no means a cure all. They could yet be squeezed out as competition to establish poverty-reducing best practices intensifies, or face considerable growth constraints if the evident disincentives to heightened levels of responsibility in Africa remain unchecked. A case is also made below that developments in the cotton issue area underscore the need for a more nuanced categorization of CSR types. Moving forward, analysts of these new phenomena should consider paying particular attention to the implications of private regulatory competition and look more closely at the ways that traditional philanthropic giving is supportive of the new responsibility or detracts from particular variants of it. These positions are restated after a brief introduction to the concept and a review of recent debates that provide necessary background information and impart the broader context for this contribution.

A brief introduction to corporate social responsibility

If the late John Kenneth Galbraith and his old rival Milton Friedman had lived to debate the credit crunch and global banking crisis it is probable that Friedman would have attempted to side-step the fact that his free market utopianism had fallen into disrepute. During their hypothetical encounter Friedman would likely have offered outlandish rejoinders to Galbraith's learned indignation, and uttered words to the effect that the supposed death of the 'free' market was actually an opportunity to solidify the existence of profit-maximizing firms. He might have heralded the opening that new stresses such as skyrocketing trade finance costs – the short-term credit necessary for over 90 per cent of the $14,000 billion world merchandise trade in 2007 – afforded firms to eschew corporate social dogoodery and get back to business 'basics' (Williams 2008). Nearly 40 years ago, Friedman infamously argued that so long as any given corporation stayed within the rules of the game, its sole social responsibility was to use its resources to engage in activities designed to increase profits (Friedman 1970). Had he witnessed the 2008 crisis it is likely that the intellectual leader of the Chicago School would have similarly exhorted market players to rule out costly 'non-core' actions that were rooted in systems to advance business social consciousness, an awareness he considered to be misguided and essentially unsound. Galbraith would no doubt have taken issue with this basically static view. He may well have reminded his adversary that considerable numbers of socially concerned and creative entrepreneurs had launched a dynamic Schumpeterian (Schumpeter 1950: 68) wave of 'ethical' innovation that has spilled across many industries. The towering figure of Keynesianism in the United States could have underscored the point that individuals and organizations that have adopted 'stakeholder' orientations or supplied burgeoning markets with products that have been vetted against an array of ethical criteria have recently thrived through focusing on much more than the so-called bottom line.

However, Galbraith would have been at a loss if Friedman had pressed him to predict whether these new practices, markets and associated business opportunities would continue their impressive expansion or even be viable in more tight and volatile economic times. There are no historic precedents that would enable an economist of his stature or of any other theoretical orientation or persuasion to produce an accurate projection of the impacts that a more volatile economy could have on the demand for these 'ethical' goods or the supply of CSR. These phenomena emerged during a prolonged economic expansion and represent novel forms of utility maximizing behaviour in markets. As such, it is improbable that recourse to the standard econometric toolkit would be able to establish with any certainty the degree to which they will mirror or escape general trends during the current period of deleveraging and retrenchment.

For the moment then, it must be assumed that CSR will remain a relevant topic for academic and policy-oriented analyses of globalization and poverty. In its simplest formulation, this type of business virtue can be said to exist when companies embrace practices that exceed minimum legal obligations in order to improve workplaces and benefit societies (Vogel 2005). To greater or lesser extents, CSR aims to integrate social and environmental considerations into everyday business operations. In so doing, CSR is argued to address the concerns of stakeholders within and beyond 'responsible' firms to raise standards of living, foster sustainable development and preserve profitability over the long term (Hopkins 2005). As CSR can entail deliberative multistakeholder interaction to establish new norms of corporate behaviour, it has also been described as a process of collaborative governance that attempts to absorb societal expectations, foster social learning and attend to the moral liability of corporations (Zadek 2005).

An especially strong variant of the new responsibility has been operative in collaborative endeavours to build alternative agro-food value chains that have shunned the principle of lowest cost, rejected the appropriation of nature and sought to differentiate and revalue formerly 'bulk' commodities based upon their cultural, social or environmental attributes (Raynolds et al. 2007: 37). The move from price-based competition towards a quality economy that fair trade, organic and other third-party certification initiatives seek to bring about has necessitated and been a source of considerable entrepreneurial innovation. Political scientists at the forefront of research on this topic have concluded that the latter phenomenon is one of the principal roots of business interest in strong CSR (Bernstein & Cashore 2007; Auld et al. 2008). Devotees to particular doctrines of more responsible, verifiable conduct now deploy economic, moral, practical, reputational and strategic rationales to convince potential backers to support their vision and the behavioural changes and benefits that they believe will stem from its implementation (Porter & Kramer 2006). The rapid growth in the numbers businesses that have embraced these approaches and in the value and volume of sales of products that have been verified by third parties in recent times indicates that these justifications have been influential.

Many analytical standpoints have nonetheless underscored CSR's potential downsides. Questions have been raised about an incipient imbalance between the scattershot norm building and standardization associated with ad hoc, private attempts to secure more responsible conduct, and more comprehensive public regulatory efforts (Blowfield 2007; Tallontire 2007). The grim possibility that the benefits communities across the South do reap from corporate responsibility could be offset if 'responsible' foreign or local investors shirk their tax obligations, lobby for fiscal restraints or push for new policies that socialize their costs and divert public investments from more generalized poverty reduction strategies has also been flagged (UNDP 2009; Utting 2007). Moreover, development experts have shown that Africans also have good reasons to be sceptical of the minimalist incarnations that Joseph Stiglitz (2006a: 199) has labelled 'business social responsibility' or BSR. Take, for example, Bono and the branded retailers behind Product RED™, who have rolled out an initiative involving a line of high-end products that enable rich consumers to vote with their dollars and support the Global Fund to fight AIDS, Tuberculosis and Malaria through their purchases. They have been accused of masking the social and environmental relations of production and trade that underpin poverty (Richey & Ponte 2008).[1]

RED™ might be the most prominent example of an overt attempt to maximize the set of assets known as 'brand value' and minimize corporate reputational risk through offering an apparently pro-development choice to elite consumers. Still, it is only one of many CSR types that have raised ethical flags or obscured the issue of corporate accountability (Conroy 2007: 31). For one, internal codes of conduct have in some instances reduced levels of ethical behaviour. Their high costs, voluntary nature and lack of institutional support can mean that they do not always effectively control corporate actions, or aim to secure anything more than a social license for the firm to continue to operate (Nicholls & Opal 2005: 73). Fair trade watchers have also identified an ethical tension within alternative value chains that are subject to rigorous third-party oversight (Raynolds et al. 2007). In these new markets there are individuals and groups on both the supply and demand sides that prioritize commerce, and there are others who are more inclined to uphold development ideals. Analysts have also challenged the ways that fair trade can lock in long-distance trade patterns, entrench specializations in underperforming and potentially declining primary product export industries and detract attention from pre-existing North-South power imbalances in the international development architecture that are not directly related to trade. At the extreme, one critic has articulated impediments to CSR that belie the idea that it will make corporations more answerable for their actions. Deborah Doane (2005) has argued that short-term cost considerations will necessarily trump long-term perspectives on social benefit generation when firms are cash strapped. She has also challenged the durability of any embryonic 'race to the top' in ethical best practice and questioned whether it is even possible for consumers to drive a new culture of corporate responsibility forward.[2]

CSR and cotton: An argument

In light of these limitations and also the hopes that this new discourse has fuelled, is corporate social responsibility addressing the factors that have impoverished Africa's cotton producers, and can it do so? The empirical analysis presented in this chapter demonstrates that to a certain extent CSR is having an impact and that it could continue to make a difference. However, its 'transformative potential' is both enabled and limited by factors and entities that the leading political science output on the topic has either explicitly excluded or failed to capture very well (Cashore 2002). Regarding the former, this study shows that there is a need to integrate philanthropic activities into the scholarly treatment of CSR. Foundations and the non-core business practice of giving have made several CSR initiatives in this issue area possible. They continue to foster livelihood improvements in locales where new certification schemes are operative and also where they are not. It is also evident that initial CSR theories have to a certain extent been inattentive to variables that could impede the uptake of CSR visions outside of the North, such as the cultural economy of reciprocal exchanges and patron-client relationships in which many cotton buyers and trading firms across Africa are rooted.

As regards cotton specifically, the following example reveals a further shortcoming of recent theorizations. An ongoing, civil society-driven multistakeholder process known as the Better Cotton Initiative (BCI) could potentially set the bar for 'better' or more 'responsible' cotton production at a low level. This effort, however, cannot easily be slotted within the robust ideal types of CSR that Auld et al. (2008) have articulated. It is also possible that the poverty reduction prospects for the comprehensive, deliberative and adaptive hard law-like institutions that have been termed 'non-state, market driven' governance (NSMD) have been slightly overplayed. I find a need to differentiate sub-types of this particular governance approach due to the fact that several NSMD and NSMD-like systems with demonstrably different poverty reduction potentials co-exist in the African cotton scene.[3] My research indicates that the probable poverty reducing impacts of one instance of NSMD-like governance – the Cotton Made in Africa (CMIA) labelling initiative – pale in comparison with those already realized under other more rigorous NSMD certification systems in place on various organic and fair trade certified cotton projects.[4]

While NSMD governance can be truncated and even corrupted, and CSR more broadly can function as a palliative, invite defection or suffer from a lack of coordination or alignment with public authorities and multilateral poverty reduction imperatives, the cotton case suggests that a greater supply of social and environmental protections, stakeholder rights and producer empowerment will be evident in the future (Blowfield 2007; Utting 2007). However, it is unlikely that the type of CSR set to hold sway across the continent is capable of fostering a redistribution of incomes and wealth sizeable enough to eradicate cotton poverty.

The argument commences below with an analysis of the BCI and the transnational norms for better cotton that are emerging from nongovernmental and industry inputs into this process. These voluntary and regionally tailored norms have the potential to improve conventional cotton production, but the initiative itself has been explicitly designed as a light-touch alternative to third-party certification. The following section shows the need for a more nuanced treatment of NSMD governance through attempting to differentiate between weaker and stronger variants that are operative south of the Sahara. A brief evaluation of efforts to build the CMIA label indicates that while the form of this project exudes all of the textbook characteristics of NSMD, its standards are considerably less ambitious as regards poverty reduction than other production and governance innovations of this type. To make this point, field research conducted on a comprehensive organic operation in a relatively impoverished district in Tanzania's Western Cotton Growing Area (WCGA) that found an especially strong poverty impact is then presented and discussed. The section concludes with a warning that the relative successes of the bioRe Tanzania project, while significant, should not lead activists to uphold organics as a panacea. bioRe has not yet adopted a producer-driven orientation or embraced imperatives evident in other NSMD systems, such as the cooperative form of producer organization that the certified fair trade cotton chain requires, and several evidently pro-poor outcomes on the project were generated through philanthropy and the unrelated interventions of non-governmental organizations such as Oxfam Great Britain. The final section emphasizes the culture of social 'irresponsibility' that exists within Tanzania's conventional seed cotton buying and ginning industry. It highlights the philanthropic interventions that have made a difference for some producers and their communities, and the factors that impede the uptake of CSR logic in one of Africa's most liberalized cotton markets.

'Better' cotton?

Backed by the World Wildlife Fund (WWF) and the International Finance Corporation, the Better Cotton Initiative (BCI) was launched in 2005. It aimed to realize the seemingly altruistic vision of making it possible for 'millions of farmers around the world to grow cotton in a way that is healthier for the farming community and the environment, and more economical' (BCI 2006).[5] In particular, the BCI sought to initiate a comprehensive multistakeholder dialogue that would define 'better cotton'. This process would establish global principles and criteria, and articulate mechanisms to enable these new standards to be applied through regionally specific implementation strategies and tools. As such, the Initiative essentially intended to generate an increasing supply of 'better' lint and effect an ethical differentiation in the bulk lint market. Its sponsors hoped that the BCI would leverage the commitments several branded retailers had made to source large and increasing amounts of product that could be clearly associated with enhanced environmental management practices and improved social and economic benefits.[6]

After the Initiative was prominently showcased at the 64th plenary meeting of the International Cotton Advisory Committee at Liverpool and prior to the first regional stakeholder engagements, the BCI website stated that its work would not directly address 'trade issues'. The site also characterized the road towards 'better cotton' as a capacity building exercise and asserted that the BCI process would not attempt to establish a new 'policing' mechanism.[7] A study funded by the United Nations Environmental Programme (UNEP) – a key steering committee member early on – and the Food and Agriculture Organization (FAO) released during this preparatory stage neatly elaborated the political rationales behind the latter distinction (de Man 2006). In his report on private sector involvement in the expansion of sustainable cotton production in West Africa, the consultant noted that branded garment retailers had an interest in pursuing actions that would limit the risks of non-sustainable cotton 'at the lowest possible' costs. In his view, it was likely the case that conventional cotton had been developing a 'bad' reputation over the previous years and that this notoriety was rapidly approaching the tipping point. He characterized the emergence of unfavourable public relations, such as an NGO-led campaign to link well known companies with unjust, immoral or illegal practices on African cotton farms as a clear and growing risk to supply chain security and brand value in Northern consumer markets. The study concluded that these firms were ready to support efforts to manage reputational risks so long as the transaction and control costs of any additional measures were minimized. In so doing, Reinier de Man argued that while retailers were not primarily interested in certifications that were 'as radical as organic', they could support the kind of modest, voluntary standard setting that the BCI envisaged for the bulk market.

To create better cotton a plan was executed over three distinct phases from February 2007. The first phase consisted primarily of an extensive 16-month consultation period. During this time working groups were formed and engaged in Brazil, India, Pakistan and West and Central Africa to provide inputs into the new standards and to also establish the foundations for future efforts to pilot the system. This participatory exercise led to the articulation of the first version of the global principles, criteria and enabling mechanisms for better cotton in July 2008. Public inputs on this document were then welcomed for an additional four months. A second phase commenced in mid-2008 to develop national guidance materials, indicators, implementation strategies and assessment guidelines. The regional groups informed this process and submitted pilot project proposals, and along with the BCI advisory committee, contributed to the support team's work to draft 'version 2.0' of the global standards. At the time of writing, initial piloting and field-testing were slated for the 2009–10 growing season, and the system was due to be finalized in 2010. The BCI also planned to establish a global programme to support national implementation strategies that would oversee a collection of funded, coordinated and aligned activities to promote better cotton, and its 'validity' and 'effectiveness' by that year (BCI 2008a).

Cotton growers, their non-governmental partners and cotton company officials from select West and Central African states learned of their possible roles and responsibilities at a workshop jointly hosted by the BCI and the Association des Producteurs de Coton Africains (APROCA) held at Bamako on 24–5 July 2007.[8] This workshop sought to sensitize regional stakeholders and kick-start the construction of a 'regional' working group. The project support team stressed that the better cotton system aimed to mainstream improved practices and develop intervention strategies exclusively for the production or 'on-farm' level. The team added that it would be the 'responsibility' of African producers to 'prove' that cotton grown under these new circumstances deserved better consideration and treatment in the global market (BCI 2007). The workshop seemingly deferred discussion of this matter to the future. Subsequently, it did not address the question of how exactly African growers or their associations would be empowered to promote wider knowledge of any benefits if and when they materialized.

At Bamako, participants were also told how the BCI differentiated between itself and certification schemes. The key difference between certifications and the BCI was that the latter was being designed to ensure the sustainability of conventional cotton production and was assuredly not an effort to engineer an 'alternative market'. Unlike organic certifications that prohibited the use of technologies such as chemical fertilizers and biotech seeds, better cotton was to be 'technology neutral'. As well, it would not involve the payment of any economic bonuses at the farm gate or further downstream, such as the mandatory price and social premia associated with fair trade certification. While noting a need for 'harmonization and synergies' with other approaches, team members variously described these systems as 'additional' to BCI, and portrayed better cotton as a 'complementary' or counterbalancing force. From the first consultation in Africa then, it was clear that the doctrines of producer responsibility, technological pluralism and adherence to market prices were not up for discussion. Any participatory input to the contrary was consequently invalid or futile *a priori*.[9]

A subsequent meeting of the West and Central African 'regional' working group drew attention to a stark tension underlying the Initiative that revealed its fundamental flaw as regards poverty eradication (BCI 2008b).[10] On the one hand, its backers wanted to develop an assessment system that would define both minimum and progress requirements, and utilize nationally tailored indicators on process, results and impacts. They visualized systems that were scalable and differentiated to farm size. Mechanisms for the verification and control of 'hoped-for' impacts on farming that could be described as 'efficient, credible and simple' were preferred. On the other hand, it set out to avoid 'heavy certification as practiced today', and as such, precluded efforts to ensure the 'traceability' of the product along the value chain. BCI insiders justified their rejection of a price premium and also their intention to not develop a unique label on the grounds that concrete information on product origins under a future better cotton system would simply be unavailable. Even so, they openly stated that the system would not prevent retailers from developing their own labels to publicize the concept or inventing a recognizable way to reference better cotton in their own marketing. Accordingly, the BCI shuns the supposed rigidity and costliness of vertically integrated, traceable value chains, while enabling its corporate patrons to pursue any strategy that they can dream up to harvest reputational capital from the scheme. Consequently, it can accurately be described as window dressing. It aims to foster a better impression for cotton without providing a verifiable or bona fide guarantee that production and processing methods have actually changed and enabled social and ecological 'wins'.[11]

If retailers have the right to fashion their own image for better cotton, the first draft of the global principles, criteria and enabling mechanisms makes it clear just where the responsibility lies for its production. The six principles are each worded in a way that entirely downloads responsibility onto farmers themselves (BCI 2008a). These tenets are all prefaced with a similar introductory statement that is in each instance followed by the specification of the action desired. For example, the first principle states that 'better cotton is produced by farmers who minimize the harmful impact of crop protection practices'. The principles on efficient water use and availability, soil health, natural habitat conservation, fibre quality preservation and the promotion of decent work are equally compromised. The team claims that it designed these otherwise noble instructions and the associated particulars to emphasize the aspects of improvement that are under 'the control of the farmer'. However, they did not stop to question whether the 'African' farmer presently exercises total, partial or even any 'control' over the preferred outcomes. In my analysis, the head of her household, or her landlord, community elders, patrons, district officials, regulators, buyers, lenders, input dealers or other local gatekeepers oftentimes have effective power over the potential for these principles to be realized. Adverse geography, poor socio-economic standing, status as a cultural or religious minority and a host of other factors that are out of her hands can and most likely will continue to limit her power without a redistribution of opportunities at the local level.

Having ceded global actions to reform trade or stimulate increased assistance flows to other agents of development, it would seem reasonable to assume that the BCI would target community-level asymmetries directly. Regrettably, the Initiative is set to engage in the kind of capacity building Bendaña (2006) has critiqued for attempting to contest power with platitudes. The as yet unfunded enabling mechanisms seek only to increase knowledge sharing and skills development, enhance producer organization and effectiveness, and facilitate more equitable access to responsible financial services. Any improvements to extension, education, organizational development or financial intermediation would certainly be most welcome. It cannot be taken for granted, however, that this limited vision will permit growers to appreciably increase their future 'control' over production.

In sum, this instance of BSR is unique. Non-governmental and private sector players have come together to steer the creation and implementation of a voluntary and differentiated corporate code through stakeholder dialogue and a relatively global piloting process. As such, the BCI cannot be characterized solely as an attempt by industry figures to spread the costs of uncovering and realizing new upstream practices and innovations that could build and secure the confidence of consumers in the decency and fairness of retail cotton products. NGOs have been equal partners in a 'pragmatic' effort that simultaneously aims to 'better' cotton and enable branded garment retailers to avert future image crises that could undermine their short-term profitability and possibly their long-run viability.

Owing to the relegation of sovereign authorities and national regulators to the sidelines, this hybrid case is far from a public-private partnership. Rather, it is an example of non-state, non-governmental engagement with the private sector to generate process and production norms that 'push back' against reputational risks and the perceived costliness of third-party certification systems other NGOs and market actors have embraced.[12] An unusual governance exercise, the Initiative has deployed the language of stakeholder rights and social and environmental protection even as it has actively limited the participation of states and the representation of cotton-dependent peoples. The importance of the BCI lies in the fact that it has sanctioned a basket of norms that lacks the depth or breadth to transform the global marketplace or to address poverty in an encompassing way. The as yet ambiguous capacity of its backers to ensure follow-through on these soft targets could well render the Initiative into a public relations travesty. Moreover, its lack of coordination with legitimate public power seems set to fuel governance voids instead of filling them.

Differentiating non-state, market driven governance

The leading theorists of non-state, market driven governance (NSMD) have contended that this form of CSR has a greater potential to transform markets than other CSR types. In their view, NSMD systems exist when businesses and stakeholders establish objectives and develop rules for achieving them without the active involvement of states (Auld et al. 2008). Such governance institutions seek to exercise their non-sovereign authority in a manner that reconfigures supply chains and markets around new process and production norms that aim to generate greater social and ecological benefits. To do so, they articulate standards that participating market players must adhere to, and ensure that these firms play by the new rules through developing penalties for non-conformity and mandating that they submit themselves to independent evaluations in order to ascertain their level of compliance with the programme. Otherwise known as an accredited third-party audit, certification or verification procedure, the latter aspect of NSMD governance has been regarded as its principal distinguishing feature. In Bernstein and Cashore's (2007) analysis, the potential for the strict and autonomous enforcement of these private 'laws' is superior. They can endogenously produce normative pressures that lead non-participants to redefine their interests and bring their activities into alignment with the new imperatives. From their perspective, empirical research must not lose sight of this dynamic knock-on effect of NSMD systems. They subtly call for scholars to move beyond static assessments of the relative importance of ethical norms or strategic business calculations underlying specific initiatives. They encourage researchers to reflect on the notion that the norm cascade or sequence of events that flow from NSMD governance could belie the intentions of its backers whether or not they exuded a profit-making orientation or expressed genuine social concern at the outset.

This push for thinkers to dig deeper than what Peter Newell and J.G. Frynas (2007) have referred to as the tension between CSR's use as a business tool and its utility for social development and ecological stewardship is warranted. If scholarly outputs on particular settings determine that markets are in fact being made to work in less antagonistic ways for people and the planet and these studies reveal and explicate the causal mechanisms that have facilitated the embedding of markets, then they would clearly offer theoretical and practical insights. These pertain to otherwise daunting big picture topics such as the maintenance of the world capitalist system and the betterment of the human condition. Even if these investigations found that the changes that have accompanied certain approaches have had trivial stakeholder impacts, they could point to the obstacles and possibilities for this kind of private authority to make business about much more than any given firm's drive to increase its market share or maximize short-term profits.

However, a focus on the dynamics of individual NSMD systems without an equivalent emphasis on comparison could lead analysts astray. As the cotton case demonstrates, multiple and rival forms of this type of governance have fuelled a differentiation between lint destined for the bulk market and lint that has been subjected to one of several distinct certification schemes. In my view, this reality demands not less but more attention to the roots of the differing governance models and the business versus development dualism. We must also pay attention to the prospect that emerging systems can pursue actions that affect other approaches or impede their growth and viability. When coordination is lacking and multiple sub-types are operative, non-state, market driven governance is not a path that leads exclusively towards the realization of 'win-win' solutions for all corporations and stakeholders involved in the various systems. When adherents to a well-organized and resourced approach take measures to secure the status and durability of their preferred vision, they can push others into potentially less effective, loss-generating endeavours and foster minimalist changes.

Take for example the CMIA labelling initiative. The German mail order company Otto Versand launched this project in 2004 (Derr 2007). Otto articulated a plan to develop independently verifiable ways and means to limit the environmental damage that resulted from cotton production and to arrest any role that cotton continued to play in increasing social divides on the continent. It proposed the phased development of a certification system that would culminate in a new label, and subsequently received broad stakeholder support from public entities and corporations based in Germany and beyond.[13] In their initial form, the CMIA standards for improvement and criteria for exclusion were organized and presented under the five broad categories of water use, pesticides, fertilizers, farmer incomes and education levels. The following year Dr Michael Otto set up the Aid by Trade Foundation, a private philanthropic entity designed to support the ostensible aims of CMIA backers to combat poverty and protect the environment through prohibiting actions and developing best practices under the five headline clusters (CMIA 2006). In July 2005 the Foundation for Sustainable Agriculture and Forestry, another German foundation supportive of the proposal, commissioned consultants with proven expertise on certified cottons to study the feasibility of the project and refine these targets. This process ultimately led to the selection of Bénin, Burkina Faso and Zambia as pilot sites for the new system in 2006.[14] The CMIA website boasted that these pilots could have a positive impact on as many as 90,000 smallholder farms in the latter country and on 150,000 such farms overall. These numbers thus enabled the CMIA to estimate that the lives of nearly one million people would be improved during the initial testing phase.

As these trials continued down to the spring of 2008 simultaneous efforts were made to ensure that demand for traceable, more 'ethical' cotton produced by African smallholders would be robust. To bolster demand prospects, the CMIA project built a so-called 'demand alliance' of firms that had expressed an interest in marketing the concept and a willingness to produce or stock products that had been sourced from CMIA certified cottons. Crucially, the project envisioned a more efficient and transparent value chain that did not stand apart from the prevailing trading order. The hope moving forward was that the quantities of CMIA lint produced would be sufficiently large to attract the interest of the biggest international cotton merchants. In the view of certain CMIA proponents, if realized, this eventuality would not only maximize the number of possible downstream outlets, but also overcome an historic impediment that they believed had plagued organic and fair trade lint exports from Africa. From their perspective, low volumes of these certified cottons had led to their exclusion from mainstream distribution channels and raised the logistical costs – the costs of planning, implementing and controlling the efficient flow of these goods – that their smaller or more specialized buyers had to recover in order to turn a profit. This apparent costliness could also be traced to the relatively weak capabilities of Africa's 'ethical' lint suppliers to ensure timely delivery of flexible volumes or specified qualities. CMIA insiders asserted that these adverse circumstances had effectively priced such products out of the mass market, implied that the weaknesses were insurmountable, and cast doubt on the capacity of fair trade or organics to foster social or environmental betterment on a significant scale. They were confident that the much larger tonnages that would flow from the CMIA approach would generate cost efficiencies and ensure that the benefits of cotton certification were more broadly shared.[15] For them, Cotton Made in Africa was an unparalleled opportunity to realize a 'win-win' scenario whereby conventional textile and garment makers and sellers would earn reputational capital on the one side. On the other, a critical mass of smallholders would be enabled to take advantage of control and traceability systems that could help them to overcome 'home-grown' productivity problems and inauspicious world market conditions. CMIA public relations materials bolstered this rosy portrayal by noting the finding of market research conducted for Accenture, a CMIA partner, that most German consumers were willing to pay more for this type of product. To those involved then, it appeared that the project was destined to correct an evident failure of the 'ethical' market and do little ill.

Unfortunately, despite the seemingly wide-ranging criteria for exclusion it has articulated, the CMIA has adopted a minimalist approach to certification. At first glance its operational guidelines, measures for sustainable farming and processing, and list of unacceptable practices all appear to be credible. For instance, the latter presents the most egregious grounds for exclusion under the revised headings of 'people, planet and profit'. Banned social practices include the worst forms of child labour and human trafficking, bonded or forced labour and union busting. Slashing and burning of primary forests and the illicit use of other designated or protected resources are barred, and pesticides that have been prohibited under the Stockholm Convention on persistent organic pollutants or listed in the Rotterdam Convention are forbidden. The economic dimension proscribes 'immoral transactions' in business relations, and also makes it clear that genetically modified organisms must not harm the 'opportunities' of other growers. Additionally, specific measures and targets under these three clusters are differentiated into red, amber and green levels of compliance to foster an impression of comprehensiveness and progression. Demonstrably higher levels of primary school attendance amongst participating cotton-dependent children and buyer efforts to develop micro-level HIV/AIDS plans are both examples of constructive green-level requirements. If CMIA farmers embrace water-saving technologies, or if more than 60 per cent in a given area use safe spraying techniques and a basket of soil-friendly agronomic practices they also can be green-lighted. A similar threshold can be reached if input prices and quality appraisals are found to be transparent, farmers receive payments promptly after their crop has been delivered to a certified ginnery, and the three-year average gross margin for participants is at least 20 per cent higher than for non-collaborating farmers.[16]

Despite the stated intentions, the development effectiveness of this label seems set to be compromised by an exceptionally long and lenient phase-in period and by cost considerations that ruled out an across-the-board control system. On the former, participating companies must declare at the outset that they have eliminated unacceptable practices and present a self-assessment. After the country-level launch, these firms agree to be audited by a third party after the first two years, and commit to do better than an average 'yellow' or amber rating. Management then must develop a plan to improve sustainability. If, after the plan is executed, the first audit finds the company to have met or exceeded an average yellow rating, the business must draft a second plan that details the actions needed to eliminate all of its red flags. Another third-party audit is subsequently conducted two to four years later to determine that all the reds have been eliminated. Should the firm succeed, it must present a final plan to convert any amber ratings that it received into green ones over the next two to four years. Ultimately – anywhere from six to 12 years after entry into the programme – a third external audit is undertaken to ensure that firms and their 'suppliers' have not fallen back into the red column.

Throughout this lengthy process it appears that the certification partners will be subject to one notable and potentially devastating constraint. For 'practical reasons', the CMIA rejected the idea of pursuing an 'expensive' attempt to certify all supplier businesses. Rather, verification of adherence will start at the ginning level, and auditors will not carry out systematic individual inspections or even representative samples from the hundreds of thousands of projected 'suppliers'. This scattershot approach could be characterized as under-resourcing, and it belies the CMIA project's avowed concern with ameliorating poverty and despoliation. Relative to the other certification systems described below it seems woefully deficient. Moreover, this inattentiveness is especially problematic in light of the professed lengths to which the lint-forward aspects of the chain have been prioritized. To that end, the project has made considerable resources available for the development of an Internet-supported database to monitor the distribution of these cottons to downstream markets. This preoccupation and a generalized aversion to the fixed and variable costs of more elaborate verification procedures have diverted investments from ensuring that the stock of CMIA cotton is actually and increasingly ethical towards the means of rendering flows of these products transparent and traceable.

The CMIA label cannot be described as an innovative approach to fighting cotton poverty. Several contributors to the Better Cotton Initiative's stakeholder consultations have expressed the belief that the aims of the BCI and CMIA are almost identical (BCI 2007). Both systems do not guarantee the payment of higher prices at the farm gate. In so doing, they have ceded responsibility for fostering direct progress on income-poverty. The BCI and CMIA seek only to augment farmer revenues and margins indirectly through stimulating the introduction of new ideas about production and their uptake, and pushing for input and output markets to be more grower, worker or eco-friendly. While this is good news for buyers insofar as they will not be required to pay farmers more per kilo than they otherwise would, neither system will establish a blanket premium for lint that has been produced in special African circumstances in order to bolster the prospects for producers and exporters. As a result, they have left individuals, organizations or processes external to the new systems with the unenviable task of proving to the global lint market that African cottons are worthy of commanding a higher price for reasons unrelated to the physical quality of the product. A number of BCI participants have also worried aloud about a palpable lack of information on the CMIA. This dearth is compounded by the vacuity of the language that can and has been used to summarize its aims. In the view of certain BCI insiders, CMIA is an attempt to foster partnerships, self-analysis, capacity building, appropriate resource channelling, the recapitulation of results, learning by doing, continuous improvement, self-evaluation and an audit culture.

These buzzwords are backed only by a system with a relatively deficient transformative potential that, much like the BCI, could increase the chance that participating firms avoid competitive disadvantages without fostering on the ground changes. If other producer-traders follow the lead of Dunavant and publicize their initial self-assessments, they will be able to trumpet their unverified claims, capture any resulting short-term public relations benefits and reap guaranteed reputational capital for at least two years.[17] If new firms continue to enter into the system, retailers will be enabled to piggyback on a constant source of positive PR from up the chain. As there are no mechanisms in place to ensure the equal dissemination or airtime of any findings from the first or subsequent third-party audits that fly in the face of these declarations, CMIA could afford members of its demand alliance a virtually limitless free ride. Even if partner self-assessments were occasionally found to be inapplicable across the entirety of their operations, untruthful or entirely made up, a steady stream of pronouncements from new entrants on their ethical intentions could help to offset the risk that the label might suffer a diminished reputation in final consumption markets. Partners that commit such offences would also have a relatively easy go if non-governmental watchdogs or market campaigners based in the North picked up on their wrongdoings and initiated awareness-raising or consumer boycotts.[18] The light-touch period where 'red flags' are allowable – up to six years – is long lasting, and non-compliant suppliers could easily search for unaffiliated lint buyers offering comparable prices in China, India, Southeast Asia and other locales where lint importers have not yet embraced the idea of ethical sourcing. Faced with this eventuality participating retailers could jump ship to other systems and offload responsibility for the failure onto their suppliers or CMIA officials.

On the other hand, if efforts to bolster CMIA's legitimacy succeed, Africa's organic and fair trade exporters would face an uphill battle to secure anything more than niche status. From Cashore (2002), it is clear that legitimacy has pragmatic, moral and cognitive aspects. The CMIA's demand alliance has the capacity to inform consumers, manipulate their perceptions of ethical content and bring these views into conformity with the system's limited environmental and social ends.

Any outreach that it conducts resonates with fashionable discourses on the need for a moral economy and enables CMIA suppliers to create demand for a range of price-competitive, bulk market products that supposedly advance this imperative. Attempts to embed the label in the popular consciousness foster the impression that consumers can do right at no additional expense and as such, endogenously generate the marginalization of costlier alternatives. A concerted information campaign could countervail this bleak outcome, though it remains an open question whether or not supporters of other NSMD systems have the resources to offset CMIA propaganda.

Notwithstanding any professed complementarities with organics or fair trade, Cotton Made in Africa's push for hearts and minds is a direct challenge to the idea that the realization of greater equity along the process and production chain should command a premium at the market. The more market share the CMIA captures, the greater the likelihood that certified organic and fair trade cotton sales will face a demand constraint. Lower willingness to pay and its corollary, stagnant or declining sales growth, could undercut organic and fair trade operations that have relied upon grants, credit subsidies, supplier good will or an expanding range of patrons to get by. As such, the existence of a relatively weaker variant of market transforming NSMD certification raises the prospect that it will increase the costs of inaction for conventional non-participants, as predicted by Auld et al. (2008). In addition costs will rise for entities that have been subjected to more strenuous certification processes. At worst, the progressive uptake of the CMIA could generate adverse market conditions and financial pressures that give cash-strapped organic and fair trade operators perverse incentives to shirk their respective systems, or abandon them entirely. While it would be unfounded and overly alarmist to predict the latter outcomes, they have been mentioned to drive home the point that the pursuit of non-state, market driven governance is not necessarily a unidirectional path towards more virtuous or responsible behaviour.

That being said, more rigorous variants of this governance form can have transformative effects. Organic cotton producers, operators, certifiers and buyers are not simply engaged in a market governance system that attempts to modify the principles and practices underlying conventional cotton production in order to help farmers and avert a demand crisis in final consumption markets. They are involved in a system that advances the idea that it is necessary to move beyond the application of industrial agriculture and monocultural models to ensure that farming is more sustainable and productive over the long term. In sharp contrast with CMIA, organic cotton production in Africa focuses exhaustively on micro or farm-level governance. Organic operators aim to enable individual producers to meet the new standards, and aspects of farmer livelihoods – broadly defined – are amongst their principal concerns.

Field research conducted on the bioRe Tanzania certified organic cotton operation, for example, found that this project had effectively addressed several factors of impoverishment that had routinely plagued the indigenous Sukuma agro-pastoralists. In 2007, 1700 farmers in Meatu District, Shinyanga, cultivated organic cotton under contract to bioRe, a subsidiary of the Swiss cotton yarn and textile wholesaler, Remei AG.[19] These individuals, their families and communities were able to benefit from innovative production techniques, training, hands-on monitoring, improved credit availability, access to input subsidies, a price premium and various philanthropic interventions that alleviated several environmental, social and structural causes of poverty. Previously on the operation, the introduction of a crop rotation method involving the cultivation of legumes and sesame was found to have preserved soils that were prone to quick runoff and high evapotranspiration rates (Ferrigno et al. 2005). This earlier research also established that the practice of intercropping cotton with pest-trapping crops such as sunflowers across the project area had rendered plant protection more sustainable and effective. During my visit I learned that bioRe had been attempting to secure market outlets for harvests from the principal rotation and trap crops, including beans, chickpeas, sesame and sunflower seeds. These efforts aimed to reduce producer dependence on cash incomes from cotton and bolster local food security.

The operation is a unique vehicle for poverty eradication and an especially strong one *vis-à-vis* conventional production in Tanzania. It pursues interventions that were not directly related to seed cotton production. Beyond everyday work to implement and police an internal control system and overhaul their training system to make it more producer-friendly, bioRe has provided disaster relief assistance. Sceptics at the country-level and elsewhere have linked the organic model with various weaknesses and threats, including the high costs of certification, the possibility of reduced yields, discrepancies between the premia operators pay at the farm gate and the premia they receive in export markets, market access barriers and the assumed deficiencies of the market for organic cotton. My research, however, indicates that pessimistic takes are largely unwarranted. bioRe's road to profitability, while long, has produced a demonstration effect, generated spillovers and pro-poor interactions with other intervening non-state actors, and has helped to fuel Tanzania's burgeoning organics movement. Outcomes on this project can be taken as being indicative of the potential for globalization to have a positive impact on peoples' livelihoods in remote African cotton-growing zones. Actions executed under this type of private sector 'hard law' have viably advanced environmental and social objectives and are capable of doing so in a more effectual manner than NSMD governance that does not rely upon similarly rigorous farm inspections, the payment of premia or the practice of giving.

The bioRe operation has set up an effective parallel governance structure that supersedes the regulatory framework and service provision norms evident in the conventional cotton market described above in Chapter 4. A production manager and a team of village-level extension supervisors and field officers – upwards of 50 individuals at any given time – enforce the rules and provide guidance and resources to farmers who would otherwise be reliant on only a handful of go-to people: the district cotton inspector, input distribution agent and several ward-level extensionists. To better understand the control system and third-party recommendations, many of these employees have learned or are learning English. They have also been instructed on how to avoid conflicts of interest that can skew the distribution of benefits from participation in the project. In Tanzania, this problem has often been rooted in the cultural economy of individual advancement, and it plagued input distribution schemes and buying practices under the old primary society-cooperative-monopsony system. Officers, each responsible for only 50 farmers, visit growers every two weeks. On the surface, the system that they implement and enforce seems austere. If, for example, individuals are found to possess conventional seeds or a chemical spray pump, the farm is automatically excluded from the project for that season. Farmers that choose to grow cotton on the same plot for more than three years in a row, that lend to or borrow from a conventional cotton cultivator or who fail to allow officers full access to their farm would face a similarly harsh sanction (bioRe 2006). Overseers also exercise considerable latent power through aggregating daily records and consistently updating farm history files that inform future on-farm practices. They also serve as a basis for future assessments of compliance and guides to the ability of particular producers to stay in line.

With cotton in the ground in the aftermath of a severe drought, producers told me that this seemingly invasive approach was having a positive impact at the farm gate and on total farm productivity, and that it was also building their community. Many informed me of their delight that they had been paid top prices – between 370 to 400 shillings per kilo – the previous season and that bioRe had once again ensured that they were paid immediately at the point of purchase. One woman spoke fondly of learning how to plant her crop in rows. Others seemed genuinely impressed with the yields and knock-on effects that they had realized through crop rotation, the cultivation of sunflowers and the use of organic fertilizers. Several were also happy that bioRe had extended subsidized credit to producers who were interested in modernizing their implements. Additionally, field officers noted that the training sessions they convene several times a month are a community-building exercise. One even opined that it would be good for direct overseers to move around a bit more to keep up attendance and producer interest in order to advance bioRe's agenda and also to enable participants to meet new people and develop social relationships.[20]

This operation has been less prone to market or governance failures than the conventional cotton system, though unbridled competition, regulatory deficiencies and rent-seeking associated with the latter constantly impede bioRe's poverty eradication potential. Since 1994, when the project was known as Tansales, single-buyer, contract-based cotton farming involving the monopoly provision of inputs and extension service has not been plagued by the standard problems. Gatekeepers of the input distribution system have not used their positions to enrich themselves or their kin. The provision of inputs on credit through inter-linked (input-output) contracts has ensured that credit is generally recoverable. This approach precluded the development of a culture of perpetual default (moral hazard) that can exist when farmers know that they will receive credit regardless of their abilities or intentions to repay. Typically only 8–10 per cent of bioRe farmers per season are unable to meet their obligation to sell an organic crop. However, incidences of non-conformance with the project's internal control system are not the only reason for the persistence of these default rates. Each season during the pre-harvest months, *machingas* or illegal buying agents covertly visit outgrowers and attempt to convince them they will receive a higher price if they do not sell their crop to bioRe. Several growers and bioRe staffers confirmed that this problem was ongoing in 2007 and asserted that illegal traders often rigged their scales, failed to inform the cotton inspector of their activities and evaded paying the 5 per cent district tax on seed cotton purchases.

Surprisingly, regulators and other buyers were unwilling or unable to tackle this problem for years. Efforts to control *machingas* were muted despite bioRe's membership in the Tanzania Cotton Association (TCA) and its willingness to remain a member in good standing. bioRe had even demonstrated a significant amount of goodwill each year when it made substantial contributions to the now defunct conventional input distribution system known as the Cotton Development Fund (CDF) even though it never took delivery of any CDF seeds or chemical pesticides.[21] Regrettably, playing by the rules in this manner did not induce a substantive effort to eradicate illicit practices. To add insult to injury, local government officials made an attempt during the 2006–07 growing season to impose a 0.3 per cent tax on bioRe's total turnover. Though the motivations behind this push were not transparent, several of my respondents speculated that unscrupulous rent-seeking buyers and the offer of bribes could have been at the root of it.

Beyond the real or imagined machinations of unethical competitors, endemic petty corruption downstream also consistently raised the transactions costs of transporting, holding, clearing and forwarding organic cotton. This reality reduced bioRe's capacity to hire new staff, train more farmers, subsidize inputs or pay higher premia.[22] Even after the firm's managers have completed the informal transactions necessary to ensure that their lint bales arrive safely at the port, they sometimes are compelled to leave Meatu to oversee the loading process in person. On one occasion, company officials provided dockworkers with a consistent supply of cold drinks to motivate them to fill the 40-foot containers to their true capacity and avoid the use of an unnecessary extra container. Nonetheless, the time-consuming, expensive and non-transparent business environment only hindered the realization of pro-poor outcomes and according to my informants, did not preclude them.

As bioRe looks to the future it seeks to generate intra-household and community level empowerment that goes much deeper than its previous efforts on the farm, in the training centre and at the market. The project management team wants to help farmers take greater control of efforts to realize sustainable yields across multiple crops and make cotton work better for the soil, producer health, food security and farm income. They envision a system of village-level satellite operations that is not dissimilar from the primary society model. These new entities would take responsibility for training and the enforcement of the internal control system, employ staff and field inspectors directly and assume operating expenses, including warehouse and office space costs. This approach would transfer responsibility for the future success of a proven model to producers, and enable the firm to focus on buying, disseminating knowledge of the model and fostering its uptake elsewhere.[23] bioRe staff expressed their willingness to initiate these satellites, assist with training costs at the outset and to continue to coordinate the provision of training materials and productive inputs as necessary. If these offshoots embrace participatory and democratic governance norms they could resemble the cooperative organizational form necessary for fair trade certification by the Fairtrade Labelling Organizations International (FLO). Regarding intra-household issues, Tanzanian staff told me that they were also interested in following the lead of their sister project in India. They informed me that the bioRe Foundation had provided funds for gender empowerment education and that the development of strategies to foster more balanced distributions of workloads and resources was ongoing on that project. Furthermore, my respondents hoped that they would be able to create jobs in the domestic spinning industry. While I was on site the project manager travelled to Arusha to discuss the prospect of adding value to bioRe lint onshore.

This organic operator has acted as if it has had strong incentives not only to mitigate social and environmental costs and align its activities with long-term community benefits, but also to engage in charitable and humanitarian work. Perhaps the most significant example of this orientation occurred during the 2006 drought when village-level staff reported that school attendance had dropped by up to 50 per cent and that the children that were still able to attend did not typically have any food to eat. The management team responded to these reports and held a teleconference with its parent on the topic. As a result of these interactions, the bioRe Foundation disbursed $60,000 for an emergency school lunch programme. At the height of the drought's impact, 8500 students at 15 schools in the 11 villages where bioRe was active were able to eat a traditional meal of beans and ugali for three months. The provision of school lunches complemented and enhanced the disaster relief efforts of international organizations and non-governmental organizations such as the World Food Program, Oxfam GB and others, and stood out as an exceptional private sector contribution during the crisis. Organic cotton farmers whose crops, livestock and families were especially hard hit, and even those who did not cultivate cotton across the district, benefited from the private provision of a safety net. bioRe farmers who subsequently received goats and cash through an Oxfam GB recovery project described above in Chapter 5 were able to reap what appear to me to be the most significantly positive benefits of globalization that have been available to Tanzanian cotton producers. The globalization of organic farming via foreign direct investment and of non-governmental service provision in this instance enabled a small number of farmers to gain access to the short-term means of survival, to build and diversify their assets and to continue to benefit from the long-term poverty eradication potential of organic production.

Organic start-ups in Shinyanga and Singida indicate that bioRe has also had a significant demonstration effect. In Singida, Export Promotion of Organic Products from Africa (EPOPA), a programme to scale up the production, certification and marketing of African organics funded by the Swedish International Development Cooperation Agency (SIDA) down to 2008, supported the 2006 launch of bioSustain Tanzania. The organic operator, Dr Riyaz Haider, was able to benefit from EPOPA's provision of technical assistance. AgroEco, a Dutch consultancy that won the SIDA tender to implement the advisory and training services related to organic production under EPOPA, trained bioSustain staff on all aspects of organic production.[24] The principal and secondary crops targeted were sesame and cotton. After the first harvest marketing problems undermined the idea that sesame could function viably as the principal crop, Haider then turned to cotton, his area of personal expertise. Previously, he had studied the bioRe operation and his doctoral project had focused exclusively on organic cotton.[25] To facilitate the switch Dr Haider contracted former bioRe production manager Louis Kapanda as a consultant. Through Kapanda's efforts roughly half of bioSustain's 900 contract farmers were able to cultivate cotton during the 2007–08 marketing season. The following year bioSustain produced a certified organic crop.

Prior to these developments Kapanda had launched his own successful organic project during the drought-afflicted 2006–07 marketing season. Known as the Busangwa Organic Farmer Association (BOFA), 120 contract farmers on this operation attained strikingly high yields of up to 1200 kilos per hectare. TanCert, a third-party organic certification agent for the domestic market and IMO certification subcontractor for organic exports to Europe, inspected BOFA during its first conversion year and certified the operation's compliance with its internal control system. Farm gate prices met or exceeded those that bioRe offered, and only minor problems with the adequacy of storage facilities, the use of banned pesticides and the incorrect application of pyrethrum (to combat cotton stainers) and neem (to take out bollworms) were reported. The United States African Development Fund (USADF) and its local technical assistance partner, the Centre for Sustainable Development Initiatives (CSDI), made these outcomes possible. Months after Kapanda left his position on the remote bioRe operation on good terms to spend time with his family in Shinyanga town in 2003, a USADF representative sought him out and informed him that a US-based buyer was very interested in sourcing organic cotton from Tanzania. After securing the blessing of the Busangwa village leadership, sending soils for testing to the Ukiriguru research centre to determine the appropriate seed varieties and waiting two years for approval, a \$99,884 grant was finally disbursed. BOFA's first harvest was ginned to organic standards in Shinyanga for a substantial fee and then sold to bioRe. The CSDI Director flagged the former costs, and the costs of IMO certification, training and finding new market outlets as areas for improvement. Should Kapanda draw upon his expertise and strong connections in the country-level and global organic cotton movement[26] to transcend these impediments and develop a business plan, he could tap the USADF's 100 per cent concessional five-year lending facility and make inroads at the heart of Shinyanga's cotton-growing zone.

Given the accomplishments of these three projects and the emergence of complementarities between them it is clear that there are considerable strengths and opportunities associated with organic cotton. It has proven to be technically feasible, healthier for farmers, more environmentally sound and a source of improved social wellbeing (Ratter 2004). The success of the initial investor and evidence of mutually beneficial interactions with its new 'competitors' also seems to contradict a particularly gloomy prediction in the development economics literature. Several theorists in that field have highlighted an apparent problem first-movers in poor countries generally face: the 'fact' that social returns to their investments in human capital are higher than the returns that those that invest in human capital can possibly hope to privately appropriate (see Rodrik 2007). In this view, latecomers can avoid costly one-off and recurrent investments in education and training simply by poaching people who have already gained the necessary expertise at another firm's expense. As per the prediction, human capital has flowed from the first-mover to new organic investors in Tanzania's Western Cotton Growing Area (WCGA). The latter, however, have not used their new stocks of knowledge to undercut bioRe's sales or to provide a significantly lower-cost product to the export market. Rather, the migration of human capital from bioRe has enabled ad hoc informal conversations amongst the firms regarding their own productive innovations, such as new methods to enhance the effectiveness of particular botanical pesticides. These interactions have sometimes stimulated the other operations to take up the innovation. Economists refer to this effect as a 'non-rival knowledge market externality' or MAR spillover. Early investors in organic cotton have also not chosen to strike human capital investments as the market for organic lint has consistently doubled in size over the past years.[27] Even if these firms had entered a market with a serious upwards demand constraint it is unlikely that cooperation or collaboration would have been entirely absent. Players in this budding industry consider the verification of adherence to a set of standards to be much more important than the realization of cost-competitiveness. There is no reason to believe that they would have allowed a push to secure greater individual shares of the market degenerate into a race to reduce margins that impeded their abilities to produce and deliver a verifiable product to buyers and final consumers in this new quality economy.

Nonetheless, organic production is associated with several physical, market and legal risks. Where high levels of chemical residues are present in the soil at the outset, as the backers of one organic trial learned in Sénégal, it can be very difficult for organic yields to rise above the 600–700 kilo per hectare level for several seasons. Even in areas where chemicals were used only sporadically, new farmers face the prospect of waiting at least two years to produce a crop that can be certified organic for export to the European market (Ferrigno et al. 2005). EU Regulation 2092/91 on the organic production of annual crops stipulates a two-year conversion period, a length of time that exposes smallholders to the risk of serious hardship without a level of oversight, training and credit provision similar to efforts that have been evident in the WCGA down to the present. The new East African Organic Standard (EAOS) stipulates only a one-year conversion period, and it is questionable whether East African states have the political will, technical capacity or resources to assert the equivalence of this standard under WTO law (EAC 2007).[28] Organic exporters also must navigate a formidable set of technological, managerial, marketing and non-price risks (EPOPA 2006). Of the former, input failure and the breakdown or inadequacy of information and communication technologies are especially relevant in remote growing areas plagued by power shortages, rough roads and climate stresses. These risks can impede supply chain management and also make it difficult to realize traceability. As well, the high costs of monitoring and marketing organics can offset the organic premium and any additional premia paid for cotton with exceptional quality attributes. Past start-ups have reported shortages of botanical pesticides, problems with the procurement and distribution of organic fertilizer, arbitrary attempts to exact product-specific taxes and instances of in-transit damage. It is plausible that future investors could face similar adversities.

Also of note here is a curious case of alleged fraud. In what I believe to be the sole example of an apparent confidence scam involving organic cotton, a foreign 'investor' allegedly duped several unlucky creditors and domestic investors in 2005. Claiming to be the first organic cotton operator in the country, his 'plans' included the start-up of an outgrower scheme involving an eye-popping 10,000 contract farmers and the creation of an organic cotton college. At an investment promotion meeting convened at the seaport city of Tanga in August that year, the 'investor' asserted that he had registered farmers in the district of Handeni to pilot the project and grow an organic crop. Though the particulars are opaque, to do so, he had apparently obtained seed financing from a Swiss agency or organization based in Dodoma, the political capital, and had also recruited a group of individual investors through personal contacts and networking. This individual told attendees at Tanga that the then President Benjamin Mkapa would launch his 'Tanzania Organic Cotton' project at a cocktail party later that month at Dar es Salaam's landmark Kilimanjaro Hotel Kempinski. During that event the 'investor' produced several pictures of cotton cultivation and declared that these had recently been taken at Handeni. He also contended that an organic crop would be harvested that September.

Two years later my interviewees confirmed that the 'investor' was no longer in the country and that his vision had not materialized. They variously asserted that the money had not been spent 'in accordance with project dictates', that there had 'been mismanagement with the money', or that he had simply 'left with all the money'. One high-level source wondered if the 'investor' had even bothered to put cotton in the ground, as in her view it would have taken a lot of time and effort to convince Handeni residents to actually plant cotton, a crop that had not been grown in that eastern district for many years. From this tale, it seems that those that choose to pursue unethical ends can exploit the growing interest in 'ethical' investments just as readily as they can take advantage of investor-interest in 'normal' projects, products or services that can be portrayed as 'quick wins'. The alleged criminal behaviour underlying this case cannot be considered a generalized threat to the organic model, but it could have an unwelcome knock-on effect if it reduces the reputational capital of the legitimate, successful and poverty-reducing operations in the Tanzanian sub-sector described above.

My informal survey of the opinions on organics held by members of the Dar es Salaam policy elite suggests that many were willing to offer qualified support for the development of this approach to responsibility, though a few did raise pointed questions about its prospects. Since 2005 the Tanzania Organic Agriculture Movement (TOAM) has lobbied for an enabling legal and policy framework for organics. These efforts seemed to me to have sensitized officials to the benefits of organic methods such as crop rotations, even though the desired impact – an explicit organics policy – had not yet been articulated. TOAM sought capacity building for registration, certification and further research as it continued to propagate knowledge of the model and endeavoured to make it easier for more buyers to take it up.[29]

Interestingly, I found that their quest had to a certain extent been frustrated by the particular resonance critical findings in the available research on organics continued to have in the policymaking community. Several insiders upheld a prediction made by researchers at Sokoine University of Agriculture years earlier that the yields of input-intensive crops such as cotton would decline after conversion. Others took a slightly different approach to the same argument and asserted that there was likely a trade-off in organics between the payment of higher prices and the realization of lower yields. These respondents were simply unaware that any short-term yield reductions that did occur in Meatu were being counteracted through careful oversight, the greater availability of services and the generation of new income streams. They did not grasp the point that organic cultivation attempts to align yields with the capacity of the soil and effectively maximize productivity at a sustainable level over the long run. A few influential respondents also reported that they were supportive of organics and 'special' marketing chains, but that they remained 'uneasy' about the market or poverty reduction potential of these 'niche' products.[30] Most did not, however, consider these threats or the admittedly 'slow and low' returns to organic investments to constitute valid rationales to abandon the model. Their desire for more information on its benefits was palpable.

From these examples it is evident that variants of non-state, market driven governance have divergent potentials to alter the context for poverty reduction. Though evidence from elsewhere suggests that much more still needs to be done to ensure that organic cotton producers remain committed to organics and can participate more fully in decisionmaking processes (Dietler & Guntern 2005), bioRe has surpassed at least one top CSR expert's threshold for making a difference. It has reduced negative environmental externalities, had a social impact and is apparently scalable, though work still needs to be done to ensure that the model could work equally well for ultra-smallholders or in areas where chemical residues are particularly high. As such, bioRe exceeds the standard Michael Conroy (2007) has deemed necessary for a positive appraisal even as it hopes to do better and empower communities to take control of the new system. Reports on attempts to further the latter objective from Kédougou, Sénégal, where the FLO certification agent FLO-Cert first verified that three cotton-producer groups had met fair trade standards on 15 October 2004, have nevertheless been mixed.[31] As such, even the most stringent private sector hard law systems have faced difficulties affecting non-income aspects of poverty. Corporate social responsibility of this intensity is consequently not a silver bullet. As I have argued, the rollout of the CMIA could generate negative spillovers that undercut the precarious foundations on which the current successes of strong NSMD rest. That being said, the organic cotton projects in Western Tanzania continue to flourish. They have introduced a new idea of what it means to do business responsibly that competes with entrenched opinions and practices that have demonstrably maintained poverty in the cotton zone. I for one hope they can survive and have a positive impact on the self-serving attitudes and individualist ways of getting things done that typify conventional approaches to buying.

Conventional cotton buying, social irresponsibility and CSR prospects in Tanzania

In the Western Cotton Growing Area many cotton buyers or their agents treat cotton growers simply as their suppliers, and this climate of opinion constitutes the limits of the possible as regards the greater uptake of CSR. Few conventional cotton farmers have even had access to buyers who have been willing to support the realization of high qualities or bigger volumes. The lucky few who have been able to enter into interlinked contracts with a buyer have typically only done so for short periods of time. While a handful of outgrower schemes were tri-alled on limited bases after the sub-sector was 'liberalized', these were discontinued in short order owing to the lack of credit recoverability the aforementioned *machinga* problem ensured. The Gatsby Charitable Foundation stepped into this void in September 2007 and is attempting to pilot contract-based farming in particular locales throughout the WCGA as part of its three-year, £3 million programme to boost yields and support value addition downstream. Gatsby's success hinges on the erad-ication of unregistered, illegal buying and the cost-reducing and tax-evading logics upon which this practice and other illiberal behaviours rest.

The Foundation's mission to introduce greater responsibility through philanthropy is a complicated one. The existence of phantom buyers has given even established, long-term players an excuse to avoid upstream investments and fostered the idea that it is the 'government's job' to back any future scheme to procure and distribute inputs. As one trader-ginner told me, it

> is not our corporate responsibility to provide inputs to people. It is not the meaning of investment for me to assure supply. I borrow, pay back the banks and pay the farmer for the cotton. I shouldn't have to worry about this input shortfall!

Despite the Ministry of Agriculture's policy to increase private sector responsibility for input procurement and distribution and to exit from further involvement with crop development funds such as the CDF described in Chapter 4, ginners have pushed hard to renew state involvement and subsidization. One former CDF insider considered government participation in any future input scheme to be essential in order to counterbalance increasingly organized producer voices. For him, moves to include the government and exclude farmer representatives from the governance arrangements of any new fund would eliminate the potential for government-farmer 'collusion' to raise the costs of doing business. In his view, ginners should 'not take too hard a hit to their margins' for supporting 'cotton suppliers', and the government must work to offset the 'contributions' businesses make to crop development.[32]

Ginners are not only interested in offloading responsibility for supply maintenance and enhancement. Several have also pushed for the establishment of a limited system of effective fiefdoms around their operations. These individuals have encouraged other members of the TCA to recognize the benefits that would flow from the establishment of exclusive buying zones proximate to their ginneries. They have lobbied the cotton board (TCB) to take note of efficiencies that could be realized if ginners were given a 50 kilometre 'zoning radius', such as reduced transport costs and greater levels of transparency at the point of purchase. One supporter even asserted during an interview that the creation of local monopsonies would enable buyers to pay higher prices at the farm gate. In seeking to effectively recreate the single-buyer environment that existed prior to formal liberalization these ginners have come up against a board that remains suspicious of their intentions. Regulators continue to believe that this proposal would unduly reduce competition, and that there is a considerable risk that producer prices could fall after it was implemented. Evidence from remote areas where monopsony conditions persist suggests that it is not far-fetched to think that farmers could be taxed anew under a zoning system. In 2006–07, the sole registered buyer in one district failed to offer farmers anywhere near the price that prevailed under competitive conditions elsewhere. He paid sellers 240 TZS per kilo while producers in other parts of the WCGA received up to 420 TZS. When the head of the Tanzania Cotton Growers Association (TACOGA) learned of the situation he wrote a letter to encourage the buyer to revise prices upwards for his 'own sake', and copied this correspondence to the board and various district officials (TACOGA 2006).

The fact that major players in this supposedly hyper-liberalized sub-sector have consistently rejected the board's calls for a regulated auction system also demonstrates that they have not had an interest in realizing the ideal of fairer or more responsible buying practices. In the board's view, transparent and monitored auctions at one site per village would end the current ad hoc and profuse distribution of seed cotton buying posts that has raised quality, price and competition concerns. On the former, buying agents have been able to take advantage of producers at far-flung points of purchase since quality grading was reintroduced. Several interviewees suggested that buying agents have graded 'A' cotton as 'B' cotton, paid farmers the lower 'B' price and transported cotton to ginneries where inspectors have categorized it as 'A'. This practice has given farmers a perverse incentive to bring dirty, wet or otherwise weighted down cotton to market. Where oversight has been ineffective or non-existent agents have also been able to profit from the reality that farmers will sell for a lower price if they are paid immediately. For example, if the going rate was 300 TZS and an unscrupulous agent claimed that he only had enough cash on hand to make spot payments at 250 TZS, in all likelihood the transaction would go ahead. Under those circumstances an agent that had been furnished with a cash advance could pocket the difference immediately. Others could capture the difference if their counter-party paid out at the quoted rate after taking delivery at the ginnery. I received reports that agents had colluded informally to enable each other to engage in this practice and also learned that TCA members had made broader if ineffective efforts to coordinate oligopsony-type pricing arrangements to keep prices low during the previous seasons.

Instead of auctions, major ginners expressed their preference for the TCA (with the input of the Board) to be empowered to set the 'minimum buying price' on a fortnightly basis. How accurately such a price would reflect scarcity conditions is highly debatable in a context where the potential price-makers have demonstrated their interest in minimizing cotton procurement costs through outsourcing[33] and attempted price fixing. Many buyers portrayed price regulation as good news for growers even as they deplored the perceived high costs of any auction system that led to the payment of higher prices at the farm gate. In their view, auctions could reduce the rent-seeking opportunities buying agents enjoyed and therefore give these contractors an incentive to demand higher fees. They also highlighted the risk that established ginners with the rated capacity and downstream outlets necessary to compete on volume would win vast quantities at auction. This hypothetical winner take all outcome would nullify any informal or formal arrangement greedy or cost-conscious firms had devised to rig bids at low levels. All in all, players in Tanzania's 'liberalized' market are well on the way towards repeating the old story that the merits of free market competition are more often preached than put into practice.

Several ginners also asserted that other firms and their agents often attempted to evade taxes and avoid their contributions to the crop development fund. Under the old CDF system buyers were required to issue farmers with receipts at the time of the transaction and affix these to their passbooks, the official record of the sales individual growers had made and the seed and pesticide quantities that were owed to them. This regulation aimed to create evidence of seed cotton sales volumes that could be cross-checked with volumes recorded at the ginnery to confirm the balance ginners owed to the district for payment of the district tax on seed cotton purchases. However, deficient regulatory capacity and petty corruption enabled some buyers to under-declare the volumes they purchased. A few simply did not include receipts with their passbooks. Others provided farmers with receipts that did not document the true volumes that had been procured. At the sketchier ginneries, temporary employees hired by the cotton board's zonal director to monitor and record deliveries were induced to 'disappear' data and look the other way when unregistered trucks arrived and made phantom deliveries. According to one ginner, these 'unclean' practices were pervasive, and those close to the borders with Uganda and Burundi had an easier go. There, ginners who under-reported purchases did not even have to worry about processing costs and the possibility that their dodge could be detected later along the chain. They simply paid for the unrecorded seed cotton volumes to be transported to the border, ensured that the right payoffs were made and sold the contraband to willing buyers in Bujumbura, Kampala and beyond. In 2007, leading ginners claimed that none of those engaged in illegal cross-border trading had had their buying licenses revoked.[34]

When asked about the prospects for CSR in Tanzania, its uptake in the cotton sub-sector or relevance to particular conventional buying operations, my respondents offered an array of opinions that underscored the extent of the challenge its proponents face. One interviewee declared that he knew

> nothing about CSR in Tanzania. There is not a lot of that going on here. When I say I am not aware of it as a generic issue, I mean it is only an issue that comes up in specific policy areas or sometimes in discussions of corruption or governance.

Others echoed the sentiment that CSR was a side issue and asserted that it detracted from the role of government or from other substantive issues such as the structure of the industry itself or from trade policy priorities. Regarding the latter, an official reacted with intense hostility when questioned about the relevance of CSR. He wondered if I really expected 'this [CSR] to mitigate the problems caused by subsidies to 5000 US cotton farmers'. Another knowledgeable and well-connected source echoed this ill will when he provided a lengthy critique of the concept from the perspective of local business. What

> I understand CSR to mean in the Western world, and I could be wrong about it, is that multinationals put on a showcase. When they extract natural resources the idea is that they leave something beneficial behind for the community, some sort of window dressing. Sometimes it is about giving a fair price or perhaps it is about cleaning up a bit… . That is the way I understand the West to understand it. If that is CSR, just stop wasting our time. It is all bullshit… . A while back some people got $250,000 USD from USAID – a drop in the bucket for them but a huge sum here – for organizing a conference to discuss CSR in the health sector. They brought in consultants and organized a two-day workshop that was held in a five star hotel. Business people in the sector came and sat there and the organizers claimed that they had a conference that mobilized and sensitized the people… it was all bullshit… . Now in the Tanzanian context, when you take [CSR] down from the global level, there is a lot of tough shit and bad luck. We are caught in a rough and unjust world. You can try to make business fair, but it just ends up undermining enterprises or entrepreneurship.

Most of the policy elites I interviewed did not agree with the assertion elaborated in the final sentence of this bleak analysis. All the same, they reiterated its implicit overall call for an approach that would generate more consequential outcomes for the domestic economy. These officials, consultants and researchers wanted 'responsible' investors to do more to transfer technology and build forward and backward linkages.

One respondent also echoed the analysis of social irresponsibility presented above when he flagged the evident underprovisioning of CSR amongst conventional cotton buyers. When prompted to specify the roots of this shortcoming, he argued that most

> think that cotton is a short-term business, and there are some that conceive it as a long-term process. The [majority] do not believe they are facing an environment of certainty... . I do not see their behaviour... cultivating good relationships. They do not seem to be into making long-term investments. In my view CSR is about long-term interests. It is not charity. It is a way of recognizing that my long-term profits depend on how I live and interact with the society around me. There should be demonstration effects and publicity. There need to be knock-on effects from people doing corporate work in the community. In my time I have seen plenty of buyers that are just interested in getting the bucks. They would probably consider that to be their social responsibility.

I later learned just how prescient this observation was when one ginner brutally summarized his attitude. This individual believed that the cotton business was simply not about

> society. After... we send money for buying we do not worry any further about things. We do not do community investments. I do this as a business. I do not do charity. I do not give anything for free... . We don't deal with organic cotton. We don't think anything of it.

On another occasion, the mere mention of CSR compelled one buyer to elaborate a derisive recrimination. He argued that the origins of underperformance across the sub-sector were 'public' rather than 'private', and fumed that

> everyone at his desk [at the Ministry] is a doctor or something and there are probably more than 50 doctors there. Are they doing the job or just sitting in A/C all the time? I am in the fields running like a mad dog. I've made a massive investment in the operation here

and I need to recover the money. I am not an expert, but it seems like the Ministry is not doing its homework.

On a more positive note, philanthropy is not entirely absent from Tanzania's conventional cotton business beyond the upstart Gatsby project, a 'fact' that potentially indicates that the current climate of irresponsibility is not immutable. Copcot Cotton Trading Ltd., a subsidiary of Paul Reinhart AG, has made numerous social investments since 1999. That year Copcot disbursed funds for improvements to the roof and drainage system of the government hospital in Geita town, Mwanza. Over the following years its charitable giving facilitated the construction of primary school classrooms at various locations throughout Geita district, and also helped to build a secondary school residence, subsidize student travel expenses and expand the hospital in town. Additionally, Copcot gave the Geita police a significant contribution towards the construction of a new station, supplied electrical equipment for the police staff quarters, paid for the prison to be wired for electricity and even installed a toilet in the district commissioner's residence. The firm categorized all of these outlays – including the three million shilling toilet ($2200 USD at January 2009 market exchange rates) – as 'community investments'. Whether or not individual expenditures reflected a genuinely public orientation or were fuelled by private motivations, by my calculations this company spent the noteworthy sum of 134 million TZS between 1999 and 2006, or about $10,000–13,000 per year on these projects. Copcot also retained ten gardeners to tend the ginnery grounds and maintain a demonstration farm, an expensive investment that few other conventional operators had been willing to make. While the firm's experiment with an outgrower scheme during this period ended in failure, its more than passing interest in the community was evident in other ways.[35]

Overall though, this orientation did not cost a lot of money. In 2006, for example, Copcot sold 7200 million TZS of lint to Reinhart. If 'community' investments that year were on par with the non-inflation adjusted average of 16.75 million TZS over the previous eight years, Copcot would have only given away 0.2 per cent of its total 2006 revenue. Still, this percentage might only be apparently low. There is simply no way of knowing the nominal costs of the undocumented 'giving' that was required to obtain licenses and permissions, preserve the dependability of agents[36] or prevent the costly holdups that gatekeepers along the chain or competitors engineered in 2006 or in any other year. My guess is that these recurrent expenditures impede the growth of above-board philanthropy and more responsible postures. These costs continue to skew flows of new investments towards outlays that reflect the firm's interest in realizing a more stable business environment more so than its avowed social concerns.

On the surface, these findings back the positive assessment mainstream agricultural experts have offered up when asked on record about the social benefits set to flow from the efforts of Reinhart and other large corporations such as Cargill and (now-defunct) Dunavant to increase their investments in Africa. In an article on the topic published in *The New York Times*, Robert Bates claimed that the new interest big players had expressed in the continent was 'magnificent news' (Zachary 2007). John Baffes explained his similarly upbeat appraisal in the same piece, noting that it was partly rooted in the fact that these investors will be in Africa for a long time, and that they were 'willing to invest a lot of money and buy in good years and bad'. From the above it is clear that at least one subsidiary in East Africa has also consistently, if imperfectly, pursued interventions that aim to advance the human condition many of its local competitors have been unwilling to replicate. However, the uniformly rosy picture these thinkers have painted for the public rests upon a piece of deductive logic – the bigger the firm, the more socially responsible and consistent the conduct – that the facts on the ground do not always support. For many years one major trader demanded the whitest cottons from its Tanzanian subsidiary, a stipulation that forced its seed cotton 'suppliers' to rely and spend relatively more on chemical pesticides than other farmers.

Another major lint trader's local branch has also been more notable for the inconsistency of its market participation than the stability of its buying operations year on year. It entered the Tanzanian market post-liberalization, withdrew and then haltingly re-entered. The facts also contradict the idea that these actors have been preordained to generate a superior stream of positive social externalities. Cargill's alleged conduct during the bumper Tanzanian harvest of 2005–06 is case in point. That year the firm did not have sufficient capacity in its Shinyanga godowns and the company decided that it needed to store seed cotton at Bibiti ginnery, the same operation that ginned bioRe Tanzania's organic lint. With no space remaining in the Bibiti godowns, Cargill allegedly proceeded to pile its purchases in the Bibiti yard. As buying continued the pile grew so big that drainage ditches were dug around it. These ditches made it impossible for other conventional buyers and for bioRe to deliver their loads or collect their lint or seeds. Cargill's cotton mountain raised others' costs and demonstrated that foreign investors can and do drop the ball. In sum, notwithstanding blanket assertions to the contrary, the ostensible benefits that experts have ascribed to foreign investment have not always been on display in Tanzania's cotton zone.

On the whole, then, it appears that conventional cotton buyers in this country have taken few steps towards embracing social respons- ibility or the practice of philanthropic giving. Many are not actively pursuing a more responsible course and have made it clear that they do not desire free competition. Ironically, buyers that hold the latter view can be categorized amongst the powerful 'anti-liberalization' forces operative in traditional export crop markets. This pervasive outlook could be added to Brian Cooksey's (2003) comprehensive list of factors that have perverted the intent of liberalization in other agricultural com- modity sub-sectors noted above in Chapter 4, including the ineffective implementation of market regulation.[37] For their part, the policy elite remains sceptical that the emergence of CSR will make a difference. There is also little evidence that members of the TCA command the resources to chart a more responsible course, or are inclined to do so. As yet, private sector associations and the Tanzania National Business Council, a forum for consultation between the public and private sectors chaired by the President, have not organized a push for firms to embrace the UN call for more sustainable, poverty-reducing business strategies (UNGC-UNDP 2004). Moreover, it is improbable that significant numbers of currently 'irresponsible' lint exporters will be preoccupied with mending their ways as financial uncertainty persists.

While I do not want to risk engaging in a lengthy and overly 'presen- tist' analysis of the impact of the 2007–09 financial crisis that would most likely be out-of-date within several months, I do feel that it is important to end this section with a warning on the short-term pres- sures that the crisis unleashed that could impede the greater uptake of social responsibility in Tanzania. In so doing I also hope to impart just how small the piece of the pie that conventional cotton farmers receive really is.

The cotton price surge of 2008 initially seemed to bode well for diversified, non-import reliant producers. However, the subsequent price collapse and the financial difficulties of conventional buyers that ensued resulted in significant downward pressure on farm gate prices that could persist without considerable regulatory interventions moving forward. During the 2008–09 marketing season conventional buyers who were not able to sell their lint on a forward basis or otherwise lock in the high prices offered during the first half of 2008 faced potential loses and even insolvency as the spot prices they received fell far short of their costs and sales were increasingly difficult to make. In December 2008 the TCB claimed that ginners had not yet sold 258,675 of the 642,935 bales that had been produced that year. One particular cause of this inflated total stood out. Several Asian importers who had entered into forward contracts when the Cotlook 'A' Index price was 77 cents per pound defaulted on their obligations after the price dropped to 45 cents.[38] Unsold volumes not only strained the ability of exporters to cover their debt service charges and pay off debts, but also stretched Tanzania's deficient storage capacity and raised quality concerns.

Farmers had been paid an average of 480 TZS per kilo at the farm gate, a price that was roughly equivalent to 39 cents during the June 2008 harvest. While this was a relatively high price, it was a price that ginners could comfortably bear had they received 77 cents per pound on the world market. At 77 cents, a hypothetical 181-kilo bale of high-quality roller ginned lint or a 210-kilo bale of saw-ginned lint would have fetched roughly $306 or $350 respectively. Farmers might have been paid between $185 and $200 if top-quality seed cotton had been ginned into the lint contained in either of those bales, and a little less if they had been filled with lint derived from inferior seed cotton grades. Importantly, in most areas of the cotton zone it is likely that farmers sold an amount of seed cotton well below the level required to produce one bale of lint. Where farm sizes were small and yields remained low in Mwanza and elsewhere, cotton producers sold on average perhaps less than half the quantity necessary for one bale and earned well below the shilling equivalent of $100.

With the world price at 45 cents per pound in the midst of the crisis, lint exporters could only hope to receive between $179 and $208 per bale. As the low world price held it seemed set to place downward pressure on farm gate prices in 2009. However, offers at the farm gate did not get as low as 250 TZS per kilo, a level that would have reduced raw material costs per bale to less than $100. The world price actually climbed upwards through 2009 and eventually hit the 80 cents per pound level in early 2010. Global price volatility consequently was not a mechanism of pure income-poverty maintenance in this instance, but was a significant factor of impoverishment. Exchange rate and production trends also have not augured well for poverty reduction. During the financial crisis the depreciation of the Tanzanian shilling *vis-à-vis* the dollar raised the costs of imported inputs. Worse still, the high farm gate prices offered during the 2008 season encouraged more seed cotton production at a time when buyers were inclined to pay a lot less. Faced with looming credit problems, rising insurance costs and possible market loss, conventional buyers did not have very many incentives to pursue the so-called triple bottom line. The prospects for the greater uptake of CSR over the short to medium terms are markedly grim without external inducements, such as efforts to inform conventional firms of the opportunities and benefits associated with hardcore variants of responsibility.

Concluding remarks

On balance the evidence suggests that corporate social responsibility has had a highly uneven impact on the relationships between cotton and poverty south of the Sahara. International traders, branded retailers and players in the new 'ethical' markets have, to greater or lesser extents, demonstrated that supply chain management is not simply about cost control or efforts to realize traditional notions of physical product quality. Individuals and organizations along the value chain have acted upon new concerns for the conditions of life on Africa's cotton farms through building dedicated institutions and developing best practice norms. While the ways and means that have been deployed to ameliorate suffering and despoliation have facilitated potentially beneficial knowledge exchanges, it is also the case that all but the most rigorous forms of CSR have produced very little in terms of verifiable, on the ground changes. Robert Reich's (2007) fear that CSR could detract from the domestic policy coordination, resource mobilization and international cooperation necessary to tackle the structural causes of poverty could well be realized if the minimalist approaches enshrined in the Better Cotton and Cotton Made in Africa initiatives hold sway. In my analysis, corporate executives pushing either of these agendas deserve the 'waves of cynicism' one *Financial Times* columnist predicted would wash over CSR snake oil salespeople (Stern 2007).

However, from the successes of organics in Tanzania and early reports of pro-poor outcomes on West African certified fair trade projects, pessimism must be applied specifically and with due caution. Any overgeneralized statement on CSR risks throwing the organic or fair trade baby out with the ethical bathwater. Paul Collier (2007: 163), for one, made this mistake *in extremis* when he chastised fair trade advocates for encouraging commodity producers 'to stay doing what they are doing'. To the contrary, expert analysis backs the claims made in FLO International's promotional materials that the fair trade certification movement seeks greater equity in international trade through a range of means including the encouragement of diversification into alternative crops and the generation of local opportunities to add value (Fairtrade Foundation 2006; Raynolds et al. 2007; Conroy 2007). How well this movement is developing new trading partnerships based upon dialogue, transparency and respect on the various cotton projects that have sprung up in West Africa is an important question for future research.

At present though, there is little evidence that fair trade certification or organics have forced their way onto the international policy agenda. Fair trade supporters led a push at the 2004 UNCTAD XI São Paulo Ministerial to reassert the conference's leadership role in commodity market governance and encouraged the pursuit of studies related to food security (Nicholls & Opal 2005). Their efforts did not gain much traction. Nonetheless, a high-level UN report released in 2008 could herald a new era of policy relevance for organics. The UN report was the culmination of the work of a joint UNEP-UNCTAD (2008) capacity building task force on trade, environment and development that commenced in East Africa in 2004. It presented 15 case studies on organic agriculture and concluded that organic production can be more conducive to food security than conventional approaches.[39] Given these findings it seems that there is light ahead as regards the prospects for the strongest variants of non-state, market driven governance and responsibility to alleviate, reduce or eradicate aspects of poverty. Coupled with these findings, I hope that the continued growth of the market for organics through 2010 induces international actions that aim to globalize successful organic models.

7
Conclusions: Global Interventions and Poverty Eradication

The facts, arguments and analyses offered above belie the notion that politics and economics are distinct, necessarily separable realms of academic inquiry. I have also presented a way of knowing about the cotton problem that is very much aligned with the idea that the 'facts' of the political economic world are not independent of the methodologies, methods or languages used to investigate and describe them. In my global political economy of cotton and poverty the particular policies or practices that oppress people in specific locales and the ways and means to address these factors are to a certain extent indeterminate. As George Soros (2008: 43) has recently argued, 'perfect knowledge is not within our reach'. Like Soros, I believe that social 'reality' is a moving target and that individuals and institutions in the political economy often hold understandings and exude expectations that are necessarily subjective. From this standpoint social 'science' is an inevitably fallible exercise. Yet as this book demonstrates, I have not chosen to throw the baby entirely out with the bathwater. Striving for the truth as regards globalization and the maintenance of cotton poverty within the interstate system and the world capitalist economy seemed a useful scholarly pursuit. I have sought to add an encompassing story to the available literature at a time of growing interest in the problem. I did not think it wise to try to manipulate 'reality' through constructing a highly critical narrative that situated, questioned or rejected the existence of institutions such as the nation state, international cooperation or capitalism itself. Although this type of approach would have been academically fruitful it seemed to me a somewhat intellectually irresponsible choice. There were plenty of treatises on those topics growing dusty on the shelves while increasing numbers of academic and professional researchers, international civil servants, policymakers, development advocates and business people scrambled to understand and act upon cotton. I chose instead to try to come as close as I could to mapping the scene of these engagements and evaluating the impacts that new actors and practices associated with globalization were having on the historic relationships between cotton and poverty.

To do so I had to be much more prescriptive than the average, ostensibly disinterested scholar that maintains pretensions to 'science' might be comfortable with. Establishing a baseline for the evaluation of globalization necessitated judgement calls, exceptional vigilance as regards bias and a recognition that nobody would be well-served if I overplayed my findings. After all, the baseline itself sanctioned and legitimized certain social institutions and relations that others could quite easily consider to be instruments of poverty maintenance. As poverty is an essentially contested concept and the validity of the definition I have adopted cannot be directly verified, it is also possible that tales told by future researchers on this topic could be markedly different from my own. Their accounts might also be noticeably dissimilar and could lead to conclusions that deviate from those offered below if they adopted and deployed distinct understandings of the term globalization, another essentially contested concept. On the latter, some might find my usage of this seemingly overutilized big idea to be limited or limiting. So be it. I wish much luck to those whose work on this topic leads to different conclusions on the ways that – to borrow an expression Susan Strange often utilized – 'who gets what, when and how' might be changing in the present era.[1]

From the outset I considered my study to be both a problem-posing effort and a problem-solving exercise. Given this orientation, the final product sits uncomfortably within the critical tradition in the field of international political economy despite the unrelenting criticism of aspects of particular global interventions that I have offered up. At least one leading theorist of the critical disposition has drawn a sharp distinction between these imperatives. He has portrayed problem-solving approaches as attempts to make things work more smoothly within the prevailing order that more or less cynically posit a continuing present (Cox 1996b: 97). My hope is that I have produced an output that is a much more middling shade of grey. That being said, I have sought to understand this subject in order to change it. My historicist account of the context for poverty alleviation, reduction or eradication argued that impoverished individuals, families and communities have been constrained by conditions, including understandings of gender, at the local, domestic, regional, national, international and global levels. In so doing it drew upon diverse sources that were by no means 'perfect' or even critical. I relied upon interviews with cotton producers and policy elites, a method that can be faulted insofar as there is no way of knowing just how far particular individuals disguised their true preferences, unintentionally or intentionally distorted data or simply confused the issues at hand. I also gained insights from economic research that was not necessarily in synchrony with the cultural milieu south of the Sahara. The works of local and world-renowned economists who treated individuals as the units of analysis, deployed questionable econometric models and assumed away institutions and power relationships that I considered to be essential components of my poverty story still taught me a lot about poverty.

In the end my story exuded a problem-solving orientation when it highlighted the various structures and agents responsible for poverty maintenance and detailed the attitudinal, behavioural and policy changes required to overcome particular aspects of the affliction. I reasoned from the detailed 'facts' I had gathered to arrive at the critical and increasingly accepted general principle that liberalization of the global cotton trade in and of itself will not expedite more equitable outcomes. Along the road to understanding and directing attention to possible pro-poor solutions I also learned as John Kenneth Galbraith (1979: 93) once noted that many of the causes of poverty can also be its effects. This 'reality' made it hard for me to extrapolate specific claims from the general scene of global interventions. As a consequence, the statements regarding the poverty-reducing potentials of various non-governmental and corporate initiatives that I have made simply cannot withstand rigorous natural science-type testing. I am fully aware that any of the propositions that I offer up on these matters below could be falsified if robust counter examples are brought to light. The very point of this study – an inherently fallible exercise – has been to subject the global processes in question to critical scrutiny and challenge and to the best of my abilities ascertain the impacts. I have had to rely on anecdotes, idioms and a good deal of guesswork, and occasionally revert to controversial theories to make my points. Conclusions on this subject matter are necessarily fluid as the interactions I have studied are dynamic. Indeed, a future effort to analyse the poverty impacts of the new global phenomena in the three 'types' of African cotton sectors – competitive, concentrated and monopoly – recently articulated by Tchirley et al. (2010) could yield starkly dissimilar conclusions. If the 'facts' someday were found to have changed, as John

Maynard Keynes once remarked, I would most definitely change my mind. For now, the summary that follows reflects what I believe to be the situation.

To reiterate briefly, my research sought to build upon studies of liberalization and the introduction of competition (Cooksey 2003; Poulton et al. 2004b) through adding a global level of analysis and refocusing attention on the poverty outcomes of new strategies and practices associated with globalization. At the outset I understood globalization to be about much more than global economic interdependence and the increasing flows of finance, trade and investment that had been enabled by technological changes and new communication systems and were evident primarily in the 'core' and emerging areas of the world economy. In my view acquiescence to or unquestioning acceptance of the benefits of this process was not a litmus test for rational 'economic' thought because the phenomenon itself was highly political. I believed that the rise of deregulation, liberalization and privatization in the 'peripheral' economies, the 'reality' of heightened commodification and the spectres of cultural homogenization and ecological ruin were demonstrably political ideas, outcomes and potentials also linked to the concept. However, in my opinion the politics of intensifying and more extensive interconnections, interdependencies and transnational flows did not end there. Politics increasingly stretched across borders as new private actors and entities sought to influence state-to-state cooperation on new issues. Moreover, it seemed to me that politics was becoming truly global as growing numbers of private actors interacted to create and enter into transnational pacts or governance arrangements that were implicit or explicit forms of political authority.

Surveying this scene I wondered about the effectiveness of the development initiatives that these individuals and organizations were engaged in and their apparent global consciousness. I thought about applying these questions to the cotton problem and subsequently chose to pursue a research project that would evaluate the impacts of civil society advocacy, non-governmental service delivery and variants of CSR in this issue area. As such my approach consciously set out to augment what I perceived to be the limited conception of the forces that maintain poverty operative in the ongoing work of the World Trade Organization (WTO) to civilize the global cotton trade, and in academic and policy-oriented research on the topic. To do so an attempt was made to construct as balanced an assessment as possible of the global economic and ideational forces, and the domestic political and cultural realities, that have maintained poverty in cotton-producing zones from colonial times down to the present. To define poverty, I drew upon the emerging theoretical consensus that poverty is a multidimensional phenomenon entailing not only a lack of income, but also the deprivation of capabilities and social exclusion (Stewart et al. 2007). The output of participatory poverty assessments and interviews conducted over a six-month period of country-level research in Tanzania and Sénégal were also drawn upon to ascertain the particular relationships that have led to the absolute and relative impoverishment of cotton growers and their differential experiences of poverty.

As this facet of my work progressed I had to think about several possible points of contention regarding the nature of poverty, the lived experience of impoverishment and the dangers associated with constructing a list of prescriptions for change. I knew that I had to avoid to the extent possible the urge to infer that all cotton producers south of the Sahara were poor from the fact that some and indeed many of them were poor. Evading this error – a mistake economists have termed a fallacy of composition – was a particular challenge given that my informants occasionally made this fault or fell victim to other logical fallacies. On the latter, several interviewees made hasty generalizations about the findings from non-representative samples when they expressed the belief that cash crop producers were everywhere and always rich *vis-à-vis* subsistence or 'sustenance' farmers. There was also the hazard that in my rush to understand, analyse and help people to act better on poverty that I would perpetrate cultural injustice. As one of my respondents noted, it is possible that simply talking with and asking isolated people about poverty can introduce the notion that they are poor or enhance their feelings of being 'poor'. I learned that many already experience these feelings as a result of their exposure to the global media, their direct observations of elite consumption habits or the stories that they have heard about life beyond their village, district or region. As such, I could take some comfort in the fact that I was likely not the only culturally disruptive force that the cotton producers I interviewed had been exposed to. But realistically, I could only console myself with the hope that my interventions would in some small way lead to outputs that underscored the need for culturally respectful and vetted ways of rising above what Galbraith referred to as people's learned acquiescence or 'accommodation' to the 'equilibrium of poverty'. Like most development observers I continued to assume that the escape from chronic poverty was essentially desirable.

The distinction between absolute and relative poverty was also never far from my thoughts as the project progressed. For example, while I was drafting the manuscript the World Bank revised its poverty headcount methodology, raised the oft-criticized dollar per day measure to $1.25 and admitted that the picture on poverty progress south of the Sahara was somewhat unclear (Ravallion & Chen 2008). In light of my research these modifications seemed largely superficial. Firstly, the Bank's update did not address other failings that critics had underscored for years, such as the inadequacies of country-level survey data. It also did not deal with the 'reality' that tiny alterations in domestic poverty lines could push sizeable numbers of people above or below that line in states where the majority of citizens are relatively or absolutely income poor (Wade 2008). The headcount continued to foster the impression that access to the local purchasing power parity equivalent of $1.25 per day or more ensured that people who lived under diverse climatological, geographical and political economic conditions within the same country were not completely destitute. I had learned in the field that poverty can be differentially experienced even within the same household and that income is often only one source of impoverishment on cotton farms. Even if one were to take the Bank's income-centric understandings of absolute poverty at face value it is clear that those behind this approach have not stopped to ask if the goods that the 'poor' themselves consider necessary to avoid poverty can actually be procured with $1.25 the world over. Such an attempt might entail inordinate costs and be logistically impractical, but would only tell us about one frequently secondary feature of poverty. As my research continued I eventually concluded that an obsession with the measures and the distinction between relative and absolute poverty detracted from other issues that bear significantly on the prospects for the poor. I felt that the hands-on discovery and mapping of the multiple dimensions of cotton poverty was more practical and potentially more consequential. Down to 2010, studies of the possible price rise that could be achieved through the implementation of trade liberalization that explicitly linked the realization of higher prices with poverty reduction continued to be greeted with fawning global media attention (Reuters 2010). My hope is that the multidimensional approach developed in this book has shown beyond a reasonable doubt how problematic it is for observers and scholars to assume that trade liberalization alone can facilitate poverty reduction in Africa.

Findings

Regarding my findings the most general statement that can be made is that globalization has altered the context for poverty reduction efforts while it has raised the prospect that a growing proportion of African cotton producers will be exposed to fewer impoverishing relationships and be able to lead lives that they value more. The market fundamentalist policy framework imported from the North at the behest of donors and creditors had especially harsh impacts on producer livelihoods and downstream opportunities in countries such as Tanzania. Elsewhere, however, these effects were cushioned by slower or more nuanced uptakes of the new mantra, though the persistence of dirigisme typically entailed the preservation of state-based institutions whose operations generated poverty and fostered inequalities. As states retreated manifold individuals and organizations set out to enhance – to greater or lesser extents – the wellbeing of cotton producers at multiple field sites and governance levels and established numerous arrangements dedicated to this task. The latter activities were given impetus early in the new millennium after non-governmental figures flagged the cotton problem and launched campaigns that brought more attention to the issue than it had received prior to or during the era of structural adjustment and conditionality. Global activism and the provision of targeted technical assistance have not only made policymakers, consumers and concerned global citizens aware of particular factors of impoverishment, but have also helped traders and retailers to become cognizant of the problem and have induced several to do something about it.

Nevertheless, traction and forward movement on the spectrum of relationships that impede wellbeing is not ubiquitous. Beyond the limited emergence of 'ethical' or 'sustainable' production systems that have proven durable and directly or indirectly remedied problematic ideas and practices in input systems, on the farm and at the market, there is little evidence that we are entering into a new era of poverty-free cotton farming. The wellbeing of women, families and communities has unquestionably been improved where buyers have been more responsible and non-governmental actors have intervened to tackle directly impoverishing relationships, broader structural constraints or crises. But it is clear from my research that those who have taken delivery of these types of assistance constitute only a small minority of the tens of millions of cotton-dependent peoples south of the Sahara. As 'private' efforts to ameliorate conditions on the farm are scaled up it is likely that the stream of benefits set to flow from these initiatives will be highly constrained. If present trends hold, instead of hardcore, third-party verified approaches to certification it is probable that most farmers will have to make do with light-touch methods that have been branded pro-poor from afar, or persevere in environments where even these limited forms of responsibility are absent. One-off local non-governmental engagements and projects, and global policy activism and service delivery have all similarly failed to target anything more than aspects of the cotton problem. Non-governmental coordination at the domestic[2] and global levels, and between these levels, to arrive at and execute a comprehensive anti-poverty agenda has also been lacking down to the present. WTO rules and the lingering influence of market fundamentalist logic also ensure that advocates for government-led investments to reduce mass poverty, efforts to add more value locally or coordinate the downstream industry across African borders face a difficult task.

Against this backdrop though, the possibility that more and more farmers will face fewer obstacles is not simply an article of faith. Discourse and concrete action on portions of the baseline I established have become entrenched and are increasingly effectual. A heightened focus on governance and resource mobilization questions is evident. Cotton-specific aid is now a reality. Responsible foreign direct investment can work for the poor and even lead on to value addition opportunities. Granted, the rate of increase in the numbers of cotton farmers or cotton-dependent peoples south of the Sahara could at any given point outstrip the rate at which globalization and the global phenomena I have studied enable particular farmers to lead lives that they value more. Exogenous shocks, such as the global financial crisis or adverse climatic events could raise costs, lower incomes, reduce yields, destroy market outlets or otherwise offset or even nullify the incremental gains that lucky farmers reap through their engagements with the new global systems of responsibility. As well, there is as yet no way of even knowing whether a concerted effort to realize the counterfactual – an Africa free from the bonds of cotton cultivation – would in some ways outperform the tentative and contingent processes of amelioration presently on the move.

Given the depth and breadth of cotton poverty, in my view, the currently plodding pace of change is morally reprehensible, but by no means foreordained. The facts from Meatu in Tanzania belie any deterministic conclusion that Africa's cotton farmers are locked in the grip of an inter-state system, global capitalist economy and cultural setting that precludes their advancement. As regards the dynamic factors that bear upon poverty reduction, the parallel resource mobilization and governance structure evident on the bioRe project is proving to be incredibly effectual. The other new global phenomena I have studied also exude specific strengths and are associated with various poverty reduction opportunities. While there are also evident weaknesses and threats, taken together, the new global phenomena represent an incipient force that has the potential to transform conditions of life for the better.

My findings on corporate social responsibility and private authority draw attention to the differential capacities of new and possibly competing private regimes to develop norms and exercise the power necessary to bring about pro-poor outcomes. Despite their nominally private origins, the Better Cotton Initiative (BCI) and the Cotton Made in Africa (CMIA) product labelling scheme exude the standard criteria for the formation of 'international' regimes. They both aim to strike pareto-improving bargains or agreements to pursue behavioural changes that aspire to make African cotton producers better off without making others (including the principal backers of the initiatives) 'worse' off (Keohane 1984: 101). These approaches essentially intend to produce information or an agenda that reduces the transaction costs associated with improving the lot of the poor. If a truly comprehensive interstate regime to spell out the implicit or explicit principles, norms, rules and decisionmaking procedures (Krasner 1983: 2) were to develop in this issue area, the costs that cotton merchants might have to bear could be significantly higher than the costs associated with the pursuit and implementation of the BCI and CMIA. The Better Cotton Initiative and Cotton Made in Africa have attempted to entrench themselves as wellsprings to guide and shape reform efforts and keep the costs of 'responsibility' down. The norms and rules these transnational entities have espoused and the resultant compliance 'pull' of these new standards, prescriptions and proscriptions do not appear to me to be independent of the power and interests which created them (Finnemore & Sikkink 1998; Cutler et al. 2001).

As such, the BCI and CMIA wield structural power and discursive power. On the former, they have imposed *a priori* limits on the range of possible anti-poverty strategies for African cotton. In particular, the BCI and CMIA have set low cost agendas for action that rule out alternatives such as the universal adoption of verifiable third-party oversight systems on par with the certification and control systems evident in organics or fair trade. On the latter, they have also framed powerful new CSR narratives in this area. The BCI, for example, has demonstrated its discursive power not only through generating relatively weak norms, but also through its attempts to communicate the Initiative's ostensible complementarity with other CSR approaches. The lack of critical responses to the BCI notion that its work is aligned with possibly more substantive CSR methodologies indicates to me that the BCI at present wields significant discursive power. Recent studies of corporate power in global agrifood governance have also highlighted the importance of the direct exercise of power, and both the BCI and CMIA also seem to command this aspect of power as well (Strange 1994; Clapp & Fuchs 2009). On both the supply and the demand sides, the BCI and CMIA have exuded the relational or instrumental power necessary to get others to do things that they simply would not have otherwise done.

Clearly, the BCI and CMIA have pushed conventional cotton buyers and traders to think about and act upon aspects of their businesses they might have previously not considered. In principle I think that this aspect of the instrumental power they wield is desirable. However, I also believe that the negative consequences of their structural and discursive power offset the potential positives associated with their direct power to improve social and environmental practices and expand the market for 'ethical' cotton. Simply put, these entities could force a low-level harmonization of what it means to do 'ethical' cotton. This eventuality could relegate to the dustbin evidently more comprehensive approaches to the realization of on the ground changes.

If similar instances of transnational competition between private governance initiatives are evident in other commodity scenes, it is important for scholars in the field of international political economy to seriously consider the ramifications of these developments moving forward. Ben Richardson (2009), for example, has developed a novel framework for understanding the exercise of power and the distributional consequences of policies in the global sugar regime. Richardson's findings and other new outputs on global commodities are ripe for comparison. Another rationale for more new work on individual commodities is that it is increasingly clear that the prospects for commodity governance have shifted significantly since the last high-water mark of interest in this subject over three decades ago. During the 1970s public authorities came together at the international level to discuss aspects of the structural causes of impoverishment and asymmetries in the international commodity trade. Today, entities that do not enjoy similar levels of legitimacy or even accountability command the power necessary to leave the structural issues largely off the table and create norms that implicitly or explicitly limit the prospects for poverty eradication. The progress and pitfalls of similar attempts at collaborative self-regulation in the financial sector were laid bare as the credit crunch morphed into the global financial crisis of 2007–09. In light of the evidently wide-ranging failures of private authority underlying the financial crisis more needs to be known about prospective non-sovereign regulators in the context of development and the implications of their differences and rivalries. While individuals within one or another of these systems might profess that the relationship of their approach to others is not the least bit antagonistic, my research indicates that there are indeed divergent interests at work. The imperative is to know more about the regulatory competition that is going on beyond policymaking processes in Africa and beyond formal interstate cooperation. Non-sovereign entities currently are vying to shape what should happen as regards the production and trade of Africa's agricultural commodities, and to ultimately control the people that consume these staples. More needs to be known about these efforts.

Moving forward

The conclusions of this research are also to a certain extent at odds with the hyperbole that spilled out of the mouths of academic, business and policy leaders and into the financial press on the alleged dangers of protectionism during the financial crisis. As banks in the North took delivery of enormous taxpayer-funded subsidies in 2008, the palpable fear was that financial protectionism could intrude to limit their abilities to fund portfolio or direct investments abroad or otherwise starve their foreign subsidiaries or cut-off non-national clients. These anxieties reached a crescendo in early 2009 as several productive industries in rich countries received bailout funds and free traders latched onto the standard 'buy American' clause that had been inserted in the US fiscal stimulus package and trumpeted the dangers of a return to depression era beggar-thy-neighbour policies. With indications that the value and volume of world trade had entered into steep decline, the notion that it was economically 'irrational' to support a 'reversion' to protectionism was bandied about and held by many mainstream economic commentators to be akin to a universal truism.

As regards the international political economy of cotton it is clear that the story must be a more nuanced one. Non-governmental actors have assuredly been correct to work with poor states that depend upon cotton exports to target the particularly egregious agricultural subsidies that remain in place in the United States (Lee 2009). These efforts to challenge illiberal trade policies that unduly reduce export revenues in countries where the average GDP per capita can be below the average hourly rate that trade lawyers bill out at have nonetheless suffered from deficient relational power and have had at least one deleterious effect. African trade negotiators simply have not enjoyed the level of power or support necessary to get the United States to do what it otherwise would not do.

Even the Brazilians have had trouble on this front. Despite the WTO Appellate Body ruling against the US export guarantee programme, the Step 2 subsidy scheme and other actionable subsidies such as counter-cyclical and marketing loan payments, arbitration at the WTO to determine the level of trade retaliation that Brazil could impose remained deadlocked through 2008 (*BRIDGES* 2008e). Once Brazil was finally authorized to proceed with retaliatory measures, in March 2010 it released a list of 50 products worth $560 million that it imports from the US to be targeted for punitive duties (Politi & Wheatley 2010a, 2010b). However, there were no indications at that time that the US Congress would be willing or able to revise the offending measures out of its farm bill or that Brazil would even be able to impose retaliatory tariffs. Instead, the financial press speculated that to avoid the duties, the US would attempt to negotiate a solution involving an offer to transfer US technologies to Brazil's already industrialized cotton system. A temporary one-year deal struck in April 2010 and an agreement to extend this deal for a further two years subsequently confirmed the journalistic specu- lation. To ward off the imposition of WTO-authorized sanctions by Brazil, the US agreed to disburse $147.3 million per year until 2012 into a tech- nical assistance fund for Brazilian cotton (Beattie 2010; *BRIDGES* 2010). As such, it appeared that even the Brazilians did not command the instru- mental power necessary to effect a solution that would advance the inter- ests of the poorest cotton-exporting nations. Perversely, this 'ceasefire' in the US-Brazil upland cotton dispute, if implemented, might simply increase the capacity of the Brazilians to export more cotton, an outcome that would clearly harm the export interests of African economies that depend on cotton. Members of delegations of the Cotton Four countries to the WTO nonetheless viewed the evident 'progress' on the upland cotton file in a positive light. The Geneva-based coordinator of the C4 considered Brazil's push to have bolstered the legitimacy of the C4's demands, and he vowed to continue to pursue remedies through the Doha Round negotiations (Agazzi 2010).

While there continue to be strong rationales for the pursuit of the free trade solution, overall, the intense focus on the subsidy-induced global oversupply problem has detracted from the equally salient imperative of value addition. Although non-governmental figures in transnational networks and coalitions have pushed for trade openness, little progress has been made on achieving the levels of domestic resource mobilization, allocation and capital formation that would be necessary for African coun- tries to capture value and emulate the ways and means that presently wealthy countries and currently dominant textile and garment exporters deployed as they grew rich (Campbell 1975; Mytelka 1989; Shadlen 2005; O'Connor 2007; Chang 2008). The higher global prices associated with liberalization could improve the terms of trade of Africa's lint exporters for a time, but might be without a global price stabilizing mechanism or supply control scheme. Regardless, terms of trade improvements are not an automatic source of capital formation (Nurkse 1953: 99). To rise above a static comparative advantage bequeathed from the colonial era and 'trade down', resources would need to be significant. Exactly who in Africa could or should be asked to subsidize moves to add more value to African cotton and enter into downstream industries?

Perhaps idealistically, I posit that the elite could be tapped, and should be tapped. They could be asked to forgo consumption, and the moment is potentially propitious.

I believe that the support that members of the African Union have expressed for renewed efforts to achieve a so-called United States of Africa, if pursued with specific attention to cotton and textiles, could raise the prospects for poverty reduction. It is by no means certain that this proposal can hope to avoid vagaries of state-to-state and sub-regional relations, cultural and linguistic barriers and the personal idiosyncrasies and interpersonal rivalries that have prevented the implementation of pan-Africanism in the past. There are no guarantees that an attempt to build a 'Fortress Africa' for cotton or a broader basket of commodities could avoid the standard downside risks of an integration effort involving the realization of common external protection. Institutional and bureaucratic sclerosis might be enabled. An intra-African rent-seeking coalition could be cemented and certain exporters might feel that they have an incentive to quit the business if they detect their profitability coming under real or imagined stress. Consumer costs might also rise inordinately. The prospect that particular states or sub-regions could defect from any commitments that they did enter into to gain a temporary economic advantage would be omnipresent. Even if African leaders were united on limited but decisive points in this area it is unclear that this shared purpose would enjoy realistic, high policy resonance similar to the big idea – the need to contain an historically expansive power – that gave impetus to the common structure to govern the European Coal and Steel Community (Pinder 1998: 5). Furthermore, the viability of the entire strategy would rely on the willingness of the world to bear the costs as more trade was diverted year-on-year, and also on the abilities of the union and union members to effectively collect taxes and keep savings at home.

Yet the possible upsides remain intriguing and in my view are worth careful consideration and a little policy entrepreneurial dreaming. The formation of a customs union with a common external tariff could break down barriers that have impeded the development of intra-African input-output linkages in the past if – and this is a big 'if' – the mutual interests thesis prevails and an institutional structure can be designed to coordinate bargaining. In this setting it is possible that the trade-off between the real need to raise producer incomes and the felt need downstream to lower average costs per unit and realize economies of scale might be more transparent and lead on to more effective governance of this crucial issue. Exporters and importers in the bloc could also have greater freedom to pursue innovative counter-trade, counter-purchase or other types of barter exchanges. The community could monitor and evaluate the development effectiveness of these bargains and integration more broadly, coordinate community-level responses to finance, food or fuel crises, and present a united African front to the various transnational private authorities that seek to impose their own regulatory visions. Under a treaty to establish a cotton and textile community or a cotton and coffee community or the like, Africa might not simply have isolated organic operations or feature the odd settlement where producers enjoy fair trade certification. The entire continent could take steps to become one big fair trade village. Imagine Africa: the totally organic continent.

A pan-African attempt to emulate the levels of multistakeholder cooperation and coordination that were evident during the development and articulation of the East African Organic Standard could be one place to look for clues as to how this admittedly idealist project could get off the ground.

Vertical funds could be established to cover the costs of adjustment, development or emergencies, and as is now being discussed in Europe, with time the community itself might be able to issue 'Afri' bonds to back investments in capacity. Overall, this dream is far from a protectionist nightmare. It is compatible with several preconditions for innovation Joseph Schumpeter identified, including the free flows of new ideas, methods, organizational forms, markets and entrepreneurial spirit. It would simply apply a lesson John Kenneth Galbraith once taught regarding the 'free' market that I argue is also applicable to 'free' trade. If these 'free' things have everywhere and always been superior mechanisms of resource allocation, why then have the wealthiest countries deployed protection for a time to get where they are, and why have some of the biggest capitalist organizations replaced market relationships with internal bureaucratic hierarchies down to the present?

Regarding the phenomenon of globalization, its constituent parts and our ways of knowing about it, my research indicates amongst other things that the 'global' nature of civil society should not be overplayed. Jan Aart Scholte (2002, 2005, 2007) has shown that the political space where voluntary associations seek deliberately to shape policies, norms and deeper social structures has spilled across borders. He has produced a list of the possible benefits and costs of the activities of individuals and organizations within this space that my findings support. Non-governmental organizations have given voice to farmer concerns with falling world prices, fuelled debate on the topic at the international level and promoted more effective global governance of the cotton problem. While the cotton-related components of Oxfam International's Make Trade Fair campaign did to a certain extent reflect farmer concerns and produce outputs that identified other aspects of the cotton-poverty nexus, the campaign was for the most part 'centrally planned' (Atkinson & Scurrah 2009). Such top-down, North-South priority setting is difficult to describe as truly 'global'. However, Oxfam's good work on cotton was clearly more 'global' than the approach that another powerful NGO adopted. From its first intervention on cotton, the IDEAS Centre has pursued a flawed top-down agenda that I argue has dubiously excluded relevant policy objectives and detracted from a comprehensive global approach in this area. Consequently, it would simply be inappropriate to ascribe the term 'global' to the civil space where non-state actors engage the cotton problem. As regards cotton, we can now only speak of an imperfectly 'globalizing' civil society. It might be possible to also argue that a 'transnational' epistemic community has developed on this topic, but here too, without attention to the vastly differing levels of relational, structural and discursive power actors in this community command, such a label risks obscuring more than it reveals.

The focus on poverty that I have pursued also points to the need to revisit the idea of 'decoupling' and explore the issue of inequality moving forward. This area for future research is especially appealing in an era of globalization where the popular intellectual lodestones of globalism that bookended the 1990s – the strong one-worldism of Kenichi Ohmae (1990) and Thomas Friedman's 'golden straightjacket' (2000) – no longer seem to apply. For a time, the financial crisis of 2007–09 was seen to discredit the notion that emerging markets were somehow 'decoupled' from market forces and trends in the financial core. The crisis also revealed just how far the financial 'industry' itself has decoupled from the productive economy and generated gross global inequalities. On the latter, some back of the envelope calculations on cotton are instructive. This commodity shot on to the world trade stage as a result of the possible impacts that up to $3.9 billion per year in subsidies to 25,000 US growers were having on the world price. These subsidies directly affected the livelihoods of tens of millions of cotton farmers who typically earned only a few hundred dollars or less annually. To make a conservative estimate, there might be 30 million people south of the Sahara that depend directly or indirectly on cotton cultivation. To make a very liberal estimate, cotton dependent households might on average earn the equivalent of $1.50 per day. For the present heuristic I will make the even more unrealistically high assumption that every man, woman and child remotely connected to the industry earns $547 per year, a level that would yield $16.4 billion in annual earnings from cotton across the continent.

If this level held today, the entire population of cotton-dependent people in Africa would be surviving on an amount roughly equivalent to Time Warner's net fourth quarter loss in 2008, or to the 'taxpayer aid' that US mortgage finance company Fannie Mae claimed that it needed in early 2009 to combat deteriorating conditions in the housing market. In percentage terms, Africa's cotton-connected could expect to earn for the year the equivalent of 0.5 per cent of the *daily* turnover of $3200 billion in global foreign exchange markets. If a currency transactions tax was imposed on all currency transactions at 0.2 per cent and turnover held it would take only two and a half days to collect enough to effectively double African incomes through immediate cash transfers or launch a vertical fund for African cotton. Alternatively, to reduce this comparison to polemical absurdity, if 0.5 per cent of the proceeds from the roughly $3000 billion in government bonds that were issued in 2009 were pooled into a $16 billion cotton fund they could achieve a similar result.

However, the income-estimate for African cotton farmers I have offered up in these examples is ridiculously high. I would hazard a guess that the $16 billion figure is probably more realistically equivalent to the total financial wealth of cotton-producing households. At this level, African cotton growers and their dependents would be worth 0.04 per cent of the nominal value – $42,000 billion – of credit default swap contracts that were outstanding in 2009.

Not to belabour the point, while the paper and electronic 'wealth' creation associated with the credit derivatives bonanza and the finance boom were still in full swing in March 2007, I spoke with several farmers in Tanzania's Western Cotton Growing Area. I asked them if any people had been around to talk about cotton-farming techniques, marketing opportunities or dropped in for any other purpose. I have reproduced an indicative conversation below. Have

> any extension officers visited and taught you about cotton growing? No they have never been to our farms, we just use our own ideas. This is the first time anyone has come here to ask about cotton farming.

Have you ever spoken to anyone else about growing cotton?
No, you are the first.
Has anyone come here this year to talk to you about anything else?
No.
Does the company that buys your cotton assist you in any way?
No.

This juxtaposition, and many others that I could have presented in this story have drawn my attention to several broader, recently debated questions on inequality that I would like to address more substantively over the coming years (Kanbur 2001, 2003; Milanovic 2002, 2003; Stewart 2003, 2007; Jomo 2005; Kaplinsky 2005; Wade 2005; Sneyd 2006c; Jolly 2007; Wolf 2007). Specifically, I want to investigate whether Robert Nozick's idea that distributions of income and wealth are justified if everyone is entitled to the holdings they possess under the distribution has passed its expiry date. I would also like to construct a global level analysis of the so-called maximin principle and subject the idea to critical scrutiny that inequalities in access or distribution are justifiable only on the basis that they are beneficial for all. Similarly, I wonder if the Kuznets curve – the idea that inequality has to grow before the benefits of growth can be more broadly shared – is still relevant or is fast becoming a red herring given present trends in international and global inequality. Setting out on a mission to know if African countries are on the path towards more equitable resource distributions than those that existed prior to the colonial era, as Samuel Huntington (1968) once posited, is definitely an attractive option. So too is an attempt to apply Thorsten Veblen's insights on conspicuous consumption and pecuniary emulation to the problems of African savings.

Finally, I am interested in knowing about how progress or the lack thereof on the imperatives for action participants in The North-South Institute's multi-donor funded *Southern Perspectives on Reform of the International Development Architecture* project articulated in 2007 will bear upon inequality. Amongst other agenda items, thinkers involved with the project highlighted the challenges of policy coordination and the ramifications of aid dependence. They also honed in on a need to mitigate the structural and discursive power that abstract, econometric approaches to understanding economic development most commonly employed in US-based departments of economics seemingly command in the development knowledge industry (Sneyd 2007).

More concretely and immediately, I plan to develop this research on cotton through applying a similar methodology and transdisciplinary analysis of globalization and the political economy of poverty to the study of tropical timber in Central Africa. The core research question for this work would mirror my doctoral research question:

what impacts, if any, are factors associated with globalization such as civil society advocacy and corporate social responsibility having on the historic factors that have impoverished Cameroonians in communities that depend upon forestry?

This question opens onto a comprehensive research agenda to map the ways that the Congo Basin Forest Fund, the Green Belt Movement, Greenpeace, the Forest Stewardship Council, the Rainforest Alliance, the World Wildlife Fund and other certification initiatives and grassroots groups might be having an effect on the inhabitants of timber extractive zones. The question also makes possible an evaluation of the recent partnership of 23 timber companies known as Africa 'Wood for Life' and of the involvement of members of the Interafrican Forest Industries Association in the multistakeholder Congo Basin Forest Partnership. It will draw attention to the influence these actors are having on the work of the Commission for the Forests of Central Africa, the African Timber Organization, the International Tropical Timber Organization and the United Nations Forum on Forests (UNFF), and the ways these pressures bear upon the potential for poverty eradication. In sum, this approach will establish empirically whether or not the new ideas, institutions and practices associated with globalization under study are modifying domestic legal and regulatory frameworks and changing the status quo in a manner that makes forestry work for people.

This research programme would add a crucial comparative component to the present study. Upon completion, I would be able to engage in an empirical dialogue with academic and policy-oriented researchers on the evident similarities and differences between the lived experiences of African agriculturalists and forest-dependent people with globalization. Such a comparison would add a dynamic contribution to the body of literature on commodity or value chains and Africa in the present era. By foregrounding poverty outcomes, I hope that my future analysis of African cotton and timber production and trade can tease out the dynamic nature of global governance. The central objective moving forward is to detail more adequately the range of possibilities that Africans on the farm and in the woods aspire to, and the new obstacles that they must confront.

Appendix A: Sources for Cotton Statistics

The International Cotton Advisory Committee (ICAC) maintains the most comprehensive statistics on global cotton production, consumption, trade, stocks and price trends (ICAC 2004). Of particular note are ICAC figures that portray the declining real world cotton price down to the present. Other sources on this topic include Gabriele Baecker's (2004) research. She found the structural downward trend in this price to be roughly 0.2 per cent annually over a 40-year period. Similarly, John Baffes (2004a: 63) produced a comprehensive table that year that related the real price decline evident from 1950 to 2002. As well, an Overseas Development Institute (ODI) (2004) briefing paper also published that year included a chart that effectively conveyed the price drop over a shorter, 30-year period. This adverse trend was more pronounced – approximately 2 per cent annually – between 1994 and 2004. Louis Goreux drafted a graphic representation of this acceleration for Eric Hazard's (2005: 88) edited collection. Also in Hazard's volume, Nicolas Gergely showed that the pace of the cotton price decline had been faster than the price declines of all other agricultural commodities. The implications of this trend and its corollary for commodity exporting countries that rely upon imports of technological and capital goods have been detailed at length by UNCTAD.

Regarding Africa specifically, a presentation that is available on the ICAC website delivered by Gerald Estur at a September 2007 conference on African cotton held at Arusha, Tanzania provides a relatively current review of Africa's place in the world cotton economy (ICAC & Estur 2007). The collection edited by William Moseley and Leslie Gray (2008: 7–9) presents useful statistics on the export value of agricultural commodities south of the Sahara over time, and also on African lint production volumes and seed cotton acreages. For its part, an OECD (2006: 46) publication on cotton in West Africa includes a good chart that presents the dramatic increase in West African cotton production and dependence evident since the early 1960s. This work also contains figures on the increase in African production and trade relative to other cotton producing nations since formal political independence. Regarding Tanzania specifically, John Baffes (2004b) has produced comprehensive data on the pre- and immediate post-liberalization periods, and Colin Poulton (2006) has updated this record. To situate Tanzania in the context of other African exporters, a somewhat dated comparison of the value of Africa's cotton exports can be found in Watkins with Sul (2002). In 2001/02, Tanzania sold a value of lint roughly equivalent to two-thirds of the sales Bénin made that year. In terms of production volumes, Tanzania produced slightly more lint than Zambia and slightly less than Cameroon in 2005 (Anderson & Valenzuela 2006). The Tanzania Cotton Board regularly updates its website with information on producer prices, registered buyers and the value and volume of seed cotton purchases and lint exports (www.tancotton.co.tz).

Readers should note the numerous warnings on the veracity of statistical data that I have presented in the text. For example, it is not clear whether researchers have the capacity to accurately determine the GINI coefficient in Tanzania let alone ascertain its relevance. My queries to the Bank's country office on this particular matter were ignored.

Appendix B: Stabilization, Adjustment and Rural Livelihoods

During the 1970s the viability of the government-led approach that many non-oil exporting cotton dependent African states had embraced ostensibly to accelerate economic development was threatened. The oil price rises of 1973 impeded capital formation insofar as they redirected savings that would otherwise have been invested in capital goods imports into current expenditures and personal consumption. Fiscal mechanisms such as tax increases or other voluntary or forced savings schemes such as pension plans or limits on cash withdrawals that aimed to mobilize domestic resources for investment were generally unable to slow this trend. Objectives to devote greater proportions of windfall foreign exchange earnings from export crops that had benefited from the commodity price boom to investments in infrastructure, human capital and new technologies fell short of expectations. As the decade wore on, earnings from the export crops of principal interest south of the Sahara became more volatile and trended downward. Thereafter, the value and volume of Africa's merchandise exports virtually stagnated while the oil price and food import costs continued to climb (Toye 1994: 19; Oyejide 2004).

Faced with a low level of foreign direct investment (FDI) African governments looked ever more to the foreign market for development finance.[1] Many obtained additional and larger loans at concessional and market rates from multilateral and bilateral lenders, and also sought to expand their creditor bases. Relatively low interest rates and the recycling of petrodollars aided the latter quest by enabling hard currency loans from foreign private banks to be cheaper and more readily available. However, the decline of the US dollar and unchecked inflationary pressures undermined the ability of scaled up external resources to deliver results. Adverse balance of payments situations remained in place, savings rates were not augmented and nationalist and pan-African aspirations for the creation and promotion of value added infant industries were crushed.

Regarding the former, the dollar decline reduced the purchasing power of African currencies that had been pegged to it at high levels to facilitate imports.[2] This reality exacerbated the rising costs of basic, capital and luxury goods imports from Europe that had resulted from European price inflation. It also made it relatively more expensive for African states and firms to invest in the expansion of their export capacities due to the fact that European contractors and consultants were the principal sources of these services. Global oversupply conditions at the root of African export stagnation also ensured that these countries missed out on an opportunity that economists would have predicted. In theory, the low dollar should have helped them to increase their sales of dollar-denominated raw material exports to countries where currencies had appreciated, but this outcome did not materialize (AfDB 2008). Even in the Communauté financière d'Afrique (CFA) where the CFA franc was tied to the French franc, dollar-denominated exports stagnated and inflationary pressures were extreme.

Governments did not generally instruct their central banks to steer austere courses. As they obtained more external resources few attempts were made to impose more stringent reserve requirements or to issue enough (or any) bonds to adequately sterilize foreign exchange inflows. While interest rates were kept high in nominal terms, real rates remained low. Rather than reduce their current expenditures African governments pursued expansionary fiscal policies that fuelled inflationary pressures and continued to subsidize food and other consumables. Overall, inflation reduced the funds that were available for investments in structural transformation.[3] This permissive environment and the policies that propped it up were simply not sustainable. Domestic investment rates built up through foreign credit topped out in 1979 (Ndikumana 2000: 387).

Governments were forced to address their 'profligacy' only after the supply of cheap credit dried up and debt servicing and repayment costs rose sharply at the end of the decade. The anti-inflation policies of the US Federal Reserve under Paul A. Volcker and Bank of England after Margaret Thatcher's election triggered the African debt crisis (Toye & Toye 2004: 255). The potential for serious debt problems had been previously discussed in 1976 at Nairobi during the fourth United Nations Conference on Trade and Development Ministerial. According to John and Richard Toye, this foul possibility was given inadequate attention during the North-South debate over the means to effect a global redistribution of incomes and wealth. As North-South divides on global policy options became entrenched and the former West German Chancellor Willy Brandt led the Independent Commission on International Development Issues to broker a resolution, Africans retained their faith in African unity, global Keynesianism and political 'interventionism' (Mkandawire Olukoshi 1995). In the *Lagos Plan of Action* developed by the United Nations Economic Commission for Africa and issued by the Organization for African Unity (OAU) in 1980, African heads of state and government committed individually and collectively to promote African integration and industrialization. They desired self-sufficiency and self-reliance for the continent and believed that these outcomes could be realized within 20 years (OAU 1980; Woods 2006: 142). The new austerity rapidly curtailed their rhetorical optimism.

The second oil shock, higher financing costs and upward pressures on currency pegs compromised economic stability in many countries and raised the prospect that non-oil exporters could succumb to protracted balance of payments crises. While cotton-dependent governments fell further into arrears with the Paris Club of official donors and the informal London Club of commercial creditors the practicality of previous governance strategies was subjected to critical external scrutiny. As Ngaire Woods has detailed, one year earlier Sénégal's economy had become the test case for the International Monetary Fund's hypothesis that the region was overly reliant on an increasing stream of credit. The Government there had been compelled to agree to the Fund's first attempt to package a set of reactive policy measures together to foster economic stability and beat back inflation. The IMF adopted and promoted an approach to stabilization that made debt rescheduling and the availability of further finance conditional on the prospective recipient's adoption of macroeconomic policies such as currency devaluation, monetary and budgetary austerity and wage restraints. Instead of heeding the Economic Commission for Africa's advice to take the lead and mitigate the impact of exogenous price shocks on its African clients, the Fund chose to interject itself in African politics through attempting to limit the scope of 'interventionism'.

As regards agriculture – the principal occupation for over 80 per cent of the population south of the Sahara at the time – the World Bank (1981) subsequently thrust itself into this volatile scene with a report entitled *Accelerated Development in Sub-Saharan Africa*. The report's lead author, Elliot Berg, blamed the evident stagnation in African agriculture on a pro-industry bias in the price, tax and exchange rate regimes governments had maintained since formal independence (Gibbon et al. 1993). The *Berg Report* argued that this favouritism distorted Africa's comparative advantage in agriculture through diverting investments from rural areas. It asserted that this diversion had undermined agricultural productivity, output growth and trade and as such, had harmed producer livelihoods. In particular Berg faulted food price subsidies that relied upon the administration of artificially low farm gate prices and import rationing systems that precluded the purchase of inputs and implements he deemed necessary for rural development. He critiqued the channelling of investment dollars into high profile, capital-intensive farm projects and challenged exchange rate regimes and border restrictions that impeded agricultural trade. Berg implored African leaders to exploit low production costs and pursue policy reforms that would simultaneously enable farmers to reap gains from the international tradability of their produce and turn the domestic terms of trade in their favour.

The Bank acted upon Berg's recommendations from 1984. It brokered the idea that structural adjustments were needed to enhance the terms upon which Africans acquired and spent foreign exchange and maximize their capacities to participate in and profit from international trade (Toye 1994). As regards agriculture, adjustment packages targeted the elimination of systems that sought to administer 'artificially' low farm gate prices on pan-seasonal and pan-territorial bases. By bringing the prices paid to farmers into alignment with world prices and concurrently removing impediments to trade such as export licensing, quotas or tariffs, the Bank assumed that agriculturalists would be given an incentive to produce more and that the conditions necessary to remedy the stagnation of Africa's export volumes would be realized (Kherallah 2002). To speed the supply response governments were told to remove systems that subsidized the procurement of agricultural inputs or food prices. They were also instructed to reduce the regulatory control and influence that crop boards and parastatal enterprises exercised over agricultural markets. Proponents of these measures to liberalize and deregulate the sector upheld the view that Africa's smallholders would be the principal beneficiaries.[4]

Short-term realities of adjustment and exogenous shocks rapidly belied the opinion that reforms would have a positive impact on the poor. African governments continued to face a high risk that they would succumb to foreign exchange crises as the nominal costs of obtaining new debt and servicing the stock of existing variable-rate debt rose sharply while the bottom fell out of the world commodity market. The lack of hard currency coupled with devaluations, domestic credit tightening and other policies that official donors and creditors had imposed complicated the production and survival strategies of many agriculturalists. These people often did not have enough cash on hand at the start of each growing season to pay for imported inputs such as seeds, chemical pesticides or synthetic fertilizers. In countries that heeded the Bank's advice to remove input subsidies and deregulate input markets growers that had previously been able to gain access to seasonal credit facilities through the patronage networks that had enveloped these programmes had to adapt. They were compelled to scour their relations for alternative sources in a context of acute cash scarcity and rising import costs. Many poor people had no recourse to formal borrowing opportunities. They typically lacked collateral substitutes and could not obtain third-party guarantees at reasonable costs. The prevalence of informal patron-client relationships also made it difficult for regulated lenders to offer credit. It was hard for these entities to screen prospective borrowers, monitor their subsequent activities or threaten the loss of future borrowing opportunities (Stiglitz 1989; Poulton 1998). In many countries only the luckiest farmers that toiled on outgrower schemes and had secured iron-clad contracts for credit provision were able to avoid the yield and income shortfalls wrought by the input credit supply crunch. Sharecroppers who had understanding and flexible landlords were similarly fortunate. Beyond the few, buyers or their agents were reticent to offer farmers credit in return for the promise of exclusive crop sales in newly competitive markets. They believed that poor people would easily succumb to moral hazard and sell their crop to a competitor, and their worries were not far off the mark (Kherallah 2002).

Beyond this initial and unmitigated adjustment cost, the conditions of life on African smallholdings did not generally improve over the medium term. This bleak result held in countries that implemented rural adjustment plans and also in those that failed to reduce fiscal outlays or introduce competitive markets, cost recovery measures or other aspects of the cookie-cutter model detailed at length in a collection edited by Thandika Mkandawire and Adebayo Olukoshi (1995). The global commodity price slump ensured that real domestic producer prices did not trend upwards during the 1980s. Even where devaluations and new regulatory environments had rendered certain crops newly tradable, low world prices and persistent local inflation often cancelled out any nominal farm gate price rises that were realized. In adjusting countries, the removal of food subsidies and the maize and grain price rises that ensued made this failure all the more severe. While African smallholders faced an assortment of context-specific adversities, their status as net food grain buyers was nearly universal. Harmful local particularities included draft animal rent increases, implement price shocks, input market failures, rising patron-client transaction costs, reciprocity network breakdowns, extension service collapses, higher transportation expenses, output market malgovernance and other upward pressures on living costs such as school fee escalation and dependence upon baby milk substitutes and other consumable and durable goods imports that had become relatively more expensive. These impediments ensured that progress on poverty remained ambiguous at best despite aggregate evidence that farmers were starting to receive prices that constituted a greater proportion of the prices exporters were paid (Helleiner 1987; World Bank 2007: 98).

Analysts of the Bank's liberalizing prescriptions and the rural fallout subsequently attributed the evident underperformance to a dubious assumption at the root of the one size fits all approach, overestimations of the local capacity to respond to change and a general inattention to cultural variables. Regarding the former, a tacit understanding operative in adjustment policies was that smallholders comprised an undifferentiated mass from which uniform responses to new production incentives were expected to be forthcoming (Gibbon et al. 1993). Liberalizers did not take into account the dissimilar factors that hindered identifiable groups such as subsistence farmers, diversified agriculturalists, agro-pastoralists and smallholders that relied on monocultural production for the export market, non-agricultural employment or a combination thereof to get by. They also did not consider the fact that within these broad categories the conditions of life and income prospects more generally were quite divergent in particular places. The not uncommon reality that the lived experiences of neighbours classed as the same economic 'type' might differ significantly was similarly ignored.

Regarding overly optimistic expectations, the idea that private sector players would step up to provide services previously administered by ministries, official agencies or marketing boards was particularly off the mark (van der Hoeven & van der Kraaij 1994: 174). It was frequently not possible for small firms that had entered new markets to supply inputs or purchase crops to equal the market coverage that government entities had previously achieved due to their limited resources, tight credit conditions, the underdevelopment of capital markets and deficient transportation and communications infrastructures. These inexperienced businesses faced steep learning curves and were enveloped in dynamic contexts where foreknowledge of potential difficulties was asymmetrically held and prone to be inaccurate (Dorward et al. 1998). While new suppliers and buyers commanded some degree of market power they had limited means at their disposal to manage risks such as perishability, shifting quality standards, adverse climatic events, the collapse of backward or forward linkages and the potential effects of the latter on growers' future food and cash crop production decisions.

Advocates also misjudged the prospects for a durable supply response. Academic surveys of the after-effects of adjustment have debunked the principal rationale behind the argument that production volumes would soar post-liberalization. Feminist development economists in particular challenged the Bank's assertion that people in unique locales suffered from a blanket condition of so-called 'under-employment' and that an economic shock could unblock withheld or unrealized labour and enhance rural productivity. Their evaluations of the impact of adjustment on women – the principal brush removers, tree fellers, cultivators, harvesters, packers, transporters and household labourers – belied the disguised unemployment thesis and documented the oftentimes superhuman efforts women had to make post-liberalization to ensure the subsistence of their families, kin and communities (Meena 1991: 172). Men might have remained 'under-employed', but this reality was rooted in a cultural economy that the market fundamentalists had simply assumed away.

Gibbon et al. (1993) and Nicolas van de Walle (2001) have argued that the general cultural blindness of the Bank's adjustment-era interventions resulted in the replacement of one form of elite-driven rural wealth extraction with another. In their view, these policies entrenched asymmetrical outcomes. Despite apparently pro-market re-regulation and formal privatization political elites and the well connected were able to entrench their dominance of the countryside. After African governments melded the imported policy framework onto pre-existing informal institutions, rural rents and rent-seeking opportunities did not disappear. They flourished. From this perspective liberalization might have actually increased the numbers of supplicants for favourable government interventions or funds. As such, Africans bought into a vision that had idealistically assumed that 'perfect' markets were not only possible, but that they could be conjured into being on African soils. As the third section of Chapter 2 discusses in greater detail, Washington's blind faith that market medicine would trump local particularities proved to be grossly misguided.

Notes

Chapter 2   Historic Relationships Between Cotton and Poverty

1   Appendix B details the broader political economy of agricultural policy change across the region that commenced 30 years ago.
2   The history of cotton cultivation in Uganda was intertwined with this debate subsequent to the Buganda Agreement of 1900. This agreement empowered a new class of indigenous landlords to charge direct cotton producers rent for their small plots. Elsewhere in the colony settlers that controlled cotton plantations also continued to employ many Ugandans as labourers. After two decades of this dual system colonial officials finally came to view tenants as a cheaper source of cotton than the European planters due to the reliance of the latter on lavish subsidies (Mamdani 1996: 141). This impression was nonetheless challenged during the 1920s after tenants reduced their cotton output to protest against excessive rents. The government moved to curb the landlords and ensure production in 1928. The dispute between proponents of small-scale agriculture and those that favoured large operations outlived the colonial era and resonates to this day. See Nafziger (1986).
3   The imposition of the cotton scheme in Mozambique was an important aspect of that colonial regime's labour conscription policies and as such, was one of the factors that sparked a 1917 rebellion against conscript labour in the Zambesi Valley (Isaacman 1976).
4   To counter this well-known ploy where it occurred and to reduce payouts where it did not, buyers rigged scales to under-weigh cotton.
5   Only in rare instances under foreign rule such as during the early 1950s in British East Africa, were cotton producers able to benefit from anything approaching what today would be termed social services. Even there the anomaly was more apparent than real, and took the limited form of short-term technical education programmes that aimed to enhance land management capabilities (Rodney 1972). Families in colonial Africa perpetually bore all the risks and costs of production, including the risks of crop failure or soil exhaustion, and the high opportunity costs that time-consuming sowing, weeding, harvesting and transporting seed cotton to the point of purchase entailed when these activities impeded the husbandry of food crops.
6   Walter Rodney conceded in his seminal work that a sizeable minority of African farmers stuck with cotton production to obtain the funds necessary to acquire other goods of European origin.
7   The political pursuit of high nominal exchange rates had several downsides for rural people. Elevated pegs could function as an effective tax on agricultural raw material exports and they occasionally priced these goods out of export markets. Coupled with the low agricultural prices and relatively steep rural taxes described below, as in Asia, high fixed rates further biased the

domestic terms of trade against the countryside (Wade 2004: 76). Internationally, high valuations subsidized new urban elites when they consumed or invested abroad and were a disincentive to the rectification of the historic flow of capital from Africa to the North. This exchange rate policy was also self-defeating when it undermined the competitiveness of Africa's light-manufactured exports. However, external factors such as aid surges and commodity price spikes pressured many governments to maintain lofty pegs. For some musings on the relation of the latter issue to growth, see Dani Rodrik's blog of 17 July 2007 at http://rodrik.typepad.com/.

8 Similarly, the World Bank showed in a June 1999 *Cotton Policy Brief* that administered producer prices were on average less than one-half of their international levels under West African monopsony systems prior to the devaluation of the CFA franc in 1994. The *Brief* claimed that the gap was explained primarily by inefficiencies in the production of cottonseed oil, oilseed cakes and pellets, and also by high operating costs and other 'implicit taxes and payments'. While this report made reference to colonial legacies it did so primarily to discredit government-led, single-buyer systems and call for their thoroughgoing reform. It did not promote the idea that compensation was necessary to replace the governance model Africa had inherited.

9 Analysts that took issue with *ujamaa*-era policies showed that the allocation of governmental resources was typically skewed toward a small number of farmers. They reported on the slow pace of productivity improvements, unearthed an evident misalignment of the costs and benefits of villagization and exposed the apparent 'fact' that agricultural policy pronouncements were rarely grounded in real political processes at the local level (Hyden 2006: 121).

10 At the outset of this period cotton farmers fared better than other agriculturalists that primarily cultivated maize or rice. President Nyerere (1967: 164) highlighted their relative wealth in a 13 June 1966 speech to the National Assembly on self-reliance. In particular, he drew attention to the fact that despite a significant drop in the prices they were paid, cotton producers 'received more money than ever before in 1965' owing to their hard work to increase production volumes. The perception that cotton cultivation was relatively enriching subsequently bore less and less relation to reality as production stagnated and farm gate prices declined significantly.

11 President Nyerere was aware of the shortcomings of these societies and of the cooperative unions, the downstream organizations that were the sole seed cotton buyers and sellers of lint to the board from 1968. In calling for better and more skilled management and condemning bureaucratic dishonesty, Nyerere defended the principles of the cooperative model and argued that they were not to blame for instances where empowered individuals had exploited their neighbours (1967: 345). The cooperative unions had emerged in the early 1950s from a movement that had raised questions about the rigging of scales at cotton-buying posts and critiqued the control Asian traders exercised over the cotton trade (Saul 1973: 141). Questions about the corruption of cooperative principles within the Victoria Federation of Cooperative Unions were first raised in 1958. Allegations of inflated travel and living allowances, inexplicable honoraria and unauthorized cash advances to staff and committee members plagued the renamed Nyanza Cooperative Union through the 1960s as it grew to be the largest cooperative in Africa.

12 Informal dealings between the ruling party and the unions have persisted down to the present. In 2008, the CCM's Mwanza regional council voted unanimously to make efforts to ensure that the Government honoured an earlier commitment to offset 886 million shillings ($775,000 USD) in back pay and benefits the Nyanza Cooperative Union owed to current and former employees. The council also agreed to make additional requests in support of Nyanza's interests. It asked the Government to write-off a further 864 million owed to the CRDB Bank, the private successor to the Cooperative Rural Development Bank, and to actively resuscitate the union, formerly the largest cooperative in Africa (Kajoki 2008). Nyanza has not been able to mount cotton-buying campaigns during the past seasons and is set to shrink to 150 employees.

13 Food market liberalization had forced net food consuming, cotton-producing households to pay closer attention to their sustenance crops. Market-determined food prices reached levels during this period that were much higher in real terms than the officially controlled prices of the 1980s. It was not uncommon for rural net food consumers to devote upwards of 70 per cent of their total expenditures to food purchases (Rutasitara 2002).

14 Baffes argued that direct producers received roughly 41 per cent of export prices prior to liberalization and 51 per cent six years on. Poulton et al. (2005) surveyed this scene several years later. Their graphical representation of average seed cotton prices per kilo and the Cotlook 'A' index lint price per pound over the 1992–2004 period showed that the direct producers had not consistently secured a larger percentage of export prices. During the 2008–09 marketing season growers captured a uniquely high proportion of export prices. Their share approached the 60–65 per cent level consistently achieved in Burkina Faso, Côte d'Ivoire and Sénégal (Goreux 2004: 18). This level is unlikely to stick as the world price drops. Aggregate comparisons of nominal seed cotton and export price trends tend to obscure the reality that the most 'efficient' ginners pay farmers less per unit of their lint output than the least efficient ginners. They also do not get at another important matter: the purchasing power of the seed cotton price in remote regions and districts where the cost of essential goods and services can be relatively higher than national averages.

15 Sufficient food stores, cash on hand and on-farm storage facilities were the necessary elements of successful withholding strategies. These resources were generally available only to those whose annual sales volumes were many times over the average amount marketed. Down to the present most surveys have not stopped to ask why larger cotton producers have enjoyed superior capability endowments. Marianne Nylandsted Larsen's (2006) instructive research, for example, found that variations in seed cotton output levels 'revolve entirely' around the access her respondents had to land and to animal traction. She did not engage in any guesswork on the sources of relative enrichment operative in particular contexts. While in no way a determining factor everywhere, male networking is a relevant point of departure for further research on this important question. Men have simply had disproportionate capacities to obtain the means of capital enhancement legitimately or otherwise due to their involvement in primary societies, village governance and party activities.

16 Representatives of the Ministry of Agriculture, Cotton Board and private sector came together in 1999 to form a Cotton Development Fund (CDF). This fund aimed to use revenues from a dedicated lint levy to tender and outsource the procurement and delivery of seeds and chemicals to producers. During the 2001–02 marketing season the CDF supplied farmers with a type of pesticide that they were unfamiliar with and no resources were devoted to scaling up their knowledge of safe usage or handling practices (Shao 2002). It piloted a project to remedy the input crisis in 2002–03. This work evolved into a forced savings scheme the next season known as the passbook voucher system. Under this approach, district cotton inspectors issued passbooks to growers. These were stamped at the market to indicate the total amount sold and the shillings that had been set aside for future input purchases. In theory farmers could later claim the inputs that were owed to them at designated distribution sites. However, questionable financial management and rent-seeking along the input distribution chain to the village level ensured that only 25–32 per cent of producers took delivery (TCB 2007). As trader-ginners paid an equivalent levy to the CDF for each kilo of seed cotton that they purchased, they also had a perverse incentive to factor these costs into the farm gate prices they offered (Poulton 2006). Calls for greater accountability and a reduction of the Board's role in the CDF were aired from 2004 (Forum-Coton 2004). A March 2006 ministerial circular (URT 2006) disestablished crop funds. Cotton stakeholders subsequently created the CDF Trust, an entity that gave buyers and producers' full responsibility for inputs. The Ministry disavowed the passbook system in June 2008 (*The Citizen* 2008).

17 This composite sketch was derived from a series of individual interviews with female farmers south of Lake Victoria in March 2007.

Chapter 3   Global Trade Governance and Cotton Dependence: Beyond Poverty Maintenance

1 Intensive land use and overcultivation have in some instances reduced soil fertility. The use of inorganic agrochemicals has fuelled bioaccumulation in local ecosystems and exposed many producers to considerable health risks. Simultaneous dependence upon lint exports and food imports and uncontrolled brush clearing have also pushed carbon emissions higher.

2 Cotton-producing and consuming countries first engaged in dialogue on world stocks and production at the fourth meeting of the International Cotton Advisory Committee (ICAC) in April 1945 (Shaw 1995). Two years later while the Havana process was ongoing, ICAC members agreed that stocks were not excessive and they moved to create mechanisms to maintain the flow of information on the world cotton situation. By 1949 ICAC had a functioning Secretariat that produced statistics and made suggestions to members about potential areas of collaboration. After the ITO failed and a global oversupply of cotton was observed, the Secretariat studied the feasibility of creating a buffer stock of cotton for the purpose of price management. The establishment of an export quota system was also discussed. However, in 1954 ICAC members agreed that there was no need for measures to control the world supply of cotton.

3   As detailed in the previous chapter, governments across Sub-Saharan Africa appropriated resources from their agriculture sectors in the post-independence period. They generated rents by paying relatively low farm gate prices in single-buyer environments. These prices often constituted a small percentage of the prices parastatals and marketing boards were paid for crop exports. Consequently, it can be argued that these governments also bear some responsibility for the income-related aspects of the relative and absolute impoverishment of agricultural producers that persisted during the decades-long North-South conflict over trade and development (see Sneyd 2006a). Whether or not governments would have raised producer prices if world commodity prices were stabilized and raised is conjectural.

4   Principal amongst the bleak research outputs on the topic was Jagdish Bhagwati's (1958) work on the potential for growth in commodity-dependent countries to be immiserizing. Bhagwati warned that growth-oriented drives to scale up commodity production and export volumes could lead to gluts on global markets and lower world prices.

5   The agreement that established the Common Fund authorized the new institution to back other price management possibilities, coordinate technical cooperation and conduct research on commodities (United Nations Negotiating Conference 1980: 3). As such, the possibility that the Fund could be used to finance the formulation and management of price-increasing export quota systems or even the indexation of prices was not ruled out *a priori*. A subsequent lack of political will and insufficient finance ensured that these paper supply management options were not brought to fruition. Nonetheless, during the 1990s the Fund supported many projects to improve the technical characteristics of natural raw materials and raise their competitiveness *vis-à-vis* synthetics (Maizels 2003: 177).

6   In addition to the above, other factors that led ICAs to break down included the failure of consuming and supplying countries to participate and the inability of participating suppliers to coordinate supply (Gibbon & Ponte 2005: 49). The technical difficulty of establishing an effective threshold price for the commencement of management operations was also significant (Toye & Toye 2004). This complexity potentially undermined the effectiveness of the cocoa agreement during its early years and afflicted other under-resourced price-managing agreements.

7   The US argued that buffer stocks would be an insufficient means of supply management unless they were complimented by a system of export quotas. This was not the generally accepted view. In response to questions from members of the ICAC standing committee in 1978, then ICAC Executive Director J.C. Santley noted that the UNCTAD Secretariat had extensively documented five viable alternative methods for handling buffer stocks of cotton (ICAC 1978: 4).

8   The Institute had been established after the 20th meeting of the ICAC. At that event a decision had been made to renew ICAC's focus on the provision and exchange of information on the world market (Shaw 1995). Two-thirds of the Institute's funding came from the US, and it ceased to exist in 1994 due to a lack of participation by other producing countries.

9   Principal causes of foreign exchange shortfalls in many commodity dependent countries included reductions in demand associated with stagflation

and monetary austerity in the North, technological developments such as the advent of synthetics and overvalued exchange rates that facilitated a profusion of high-end, final consumption purchases by poor country elites abroad.

10 This declining proportion was also induced by the high average tariffs levied by Africa's developing country trading partners and by the relatively strong and highly subsidized agricultural export growth of several of those partners. Rapid population growth and the maintenance of emergency domestic stockpiles also compromised the ability of African countries to slow the decline of their historic share of world agricultural exports.

11 Including agriculture, minerals, timber, and fish but excluding oil, at least 43 countries continue to rely upon the export of three or fewer primary commodities for up to 90 per cent of their foreign exchange earnings (Watkins & Fowler 2002). Twenty-six African LDCs showed an average level of specialization on these exports of 86 per cent in 1997, a number that had not changed significantly since 1980 (Gibbon & Ponte 2005: 43). According to Gibbon and Ponte, exports of agricultural raw materials have accounted for an increasing proportion of African exports.

12 Long-term declines in the terms of trade of commodity exporting countries result principally from relatively inelastic demand for these commodities in developed countries. According to Alfred Maizels (2003: 171), real declines in commodity prices had disproportionate impacts on Africa after 1980.

13 The Agreement compelled developed countries to lower product-specific and non-product-specific support of this type by 20 per cent over six years. The WTO referred to the total figure used to calculate the 20 per cent reduction as the total aggregate measure of support (AMS). Subsidies were exempted from this reduction if they could be characterized as being directed at programmes to limit production. These measures were categorized in the 'blue' box. In any given year, if non-product-specific support was less than 5 per cent of the total value of agricultural production or product-specific support was similarly less than 5 per cent, developed countries could invoke the *de minimis* clause to exempt these measures from the mandated reduction. Developing countries were given a 10 per cent *de minimis* ceiling, and any trade-distorting investments they made or input subsidy schemes that they maintained were exempted from the amber box on developmental grounds. All signatories had agreed to reorient their systems towards the adoption of minimally trade-distorting 'green' box subsidies. Subsequent notifications to the WTO from the US and the EU on their green, amber and blue box measures indicated that the desired transfer from the amber to the green box took place over the Agreement's first six years. However, several analysts have argued that much of the evident shift was attributable to the questionable use of the *de minimis* clause loophole (Goreux 2004: 10).

14 Through 1998 the average income in Burkina Faso was 40 per cent higher in real terms than in 1970 due in part to the intensive and extensive expansion of export-oriented cotton production. Downward pressure on the world price resulting from the subsidized oversupply undermined these gains from the mid-1990s (van de Walle 2001: 278). Mats Hårsmar (2004: 174) has detailed the complex causes of increased cotton cultivation in Burkina. In his analysis, the risks posed by a declining world price were not a disincentive to the scaling up of production. Many Burkinabé producers were nonetheless aware

that their economic choice to rely upon cotton and gain access to inputs, implements and credit that would otherwise be unavailable could entail long-term costs.

15 The top 1 per cent of the nearly 25,000 US cotton farmers received about one billion dollars during the 2001–02 cotton marketing season or 25 per cent of total support (Zunckel 2005: 1076). The 255 cotton farms that were then larger than 3000 acres (1214 ha) received subsidies that averaged $1.2 million per farm in 2002–03.

16 Denis Pesche and Kako Nubukpo (2005: 48–51) have detailed the regional level precursors to the Sectoral Initiative. These antecedents included meetings at the producer and ministerial levels, and the successful dissemination of non-governmental research outputs to high-level figures.

17 The roots of this characterization and the controversy that surrounded it are elaborated fully in the discussion of non-governmental actors and the Cotton Initiative below in Chapter 5.

18 Louis Goreux (2005: 84) estimated that the annual costs of compensation would be $250 million. In his view, these costs during the 2002–03 cotton-marketing season would have been equivalent to roughly 8 per cent of the total dollar value of US and EU subsidies (Goreux 2004: 4). He argued in 2004 that compensation payments could be made to 22 African LDCs. These could be distributed according to production and export volumes and disbursed at the country level could take the form of supplementary payments made at the farm gate. Other supporters of the Initiative considered the latter idea to be an invitation to rent-seeking. They noted that lump sum payments might not trickle-down to the 10–16 million people in West and Central Africa that depended directly or indirectly on the cash incomes afforded by cotton sales (Baden 2004: 13–14).

19 The most common argument levelled against compensation was that it could give beneficiaries a perverse incentive to produce more cotton at a time when world production should have been declining (Goreux 2005: 86). African delegations countered this point and noted that aspects of the US programme, such as its counter-cyclical payments scheme, inherently generated excess supply.

20 The G90 had a unifying interest in the realization of scaled up special and differential treatment for the LDCs, including duty-free and quota-free treatment for LDC exports in the North. It also sought to achieve specific market access commitments for all developing countries, and the reduction of bound tariffs, and tariff peaks and tariff escalation on value added products (Blouin & Weston 2005). The G20's positions on agriculture articulated at Cancún resonated with this broader agenda. Contrary to all predictions, the coalition held (Narlikar & Tussie 2004).

21 Sally Baden and Eric Hazard have noted that representatives of African cotton producer organizations and non-governmental organizations were not invited to attend this workshop.

22 The report reiterated the fact that the US share of world exports increased from 23.5 per cent in 1999 to 39.9 per cent in 2002 (Zunckel 2005).

23 Writing in mid-2005 Romain Benicchio concluded that the US had to move within six months of the DSB's adoption to remove Step 2 payments. Failure to do so would have contravened Article 7.9 of the Agreement on Subsidies

and Countervailing Measures. The US spent $3.582 billion on cotton subsidies in 2005 and moved to eliminate Step 2 in July 2006 (Sumner 2007).

24 As Joseph Stiglitz (2006a: 81) noted, the offer was worth little to African LDCs as long as the US remained the world's biggest exporter and its domestic spinning and weaving industries continued to decline. The ongoing availability of subsidized cotton in Europe also reduced the potential that African countries could reap gains from this concession.

25 Rich WTO members had previously encouraged these liberalizing and privatizing reforms in their donor – or to use the current euphemism – 'partner' roles. In effect, paragraph 12 praised African members that were reliant upon policy-contingent lending for listening closely to their development 'partners' beyond the WTO and taking on board their economic advice.

26 A significant proportion of this total was directed towards marginal producers and traders of cotton lint such as Kenya. For example, the total included a $335 million PRGF disbursement to that country. Likewise, the inclusion of MDRI debt cancellation afforded to countries that had past the completion point of their Heavily Indebted Poor Country Initiative (HIPC) debt relief programmes skewed the data. For example, $582 million in MDRI extended to Zambia was counted as CDA. This categorization was made despite the fact that the cancellation of debts owed to the Bank's concessional lending arm, the IMF and the African Development Fund under the MDRI was a mechanism to reduce the stock of external debt in post-HIPC countries and not a flow of new assistance. The MDRI sought to stimulate compensatory flows only to the institutions that wrote off the odious debts.

27 The most recent version of this table prepared after the conclusion of the high-level session distinguished cotton-specific development assistance from assistance related to agriculture or infrastructure, and also from the resources that are available for cotton under the PRGF, MDRI and HIPC. In the revised table cotton-specific development assistance only totals $495 million, or 7.2 per cent of the previous draft's misleading sum (WTO 2007b).

28 The US was faulted by the panel and Appellate Body decisions for wrongly shielding trade-distorting measures in the green box. Critics charge that the slow pace of change stemmed from a concurrent US attempt to redefine the criteria used to determine blue box subsidies. A favourable redefinition would enable the USTR to shield certain offending measures in this box (Khor 2007).

29 Robert D. Putnam's (1988) research on the ways executive branches of government attempt to bridge domestic and international level negotiations is insightful on the apparent dissonance of the past administration's approach. It confronted an international imperative to change that could have had a deleterious impact on a vocal and actively hostile domestic constituency. Given its inability to align the interests of both levels it is possible that the administration offered rhetorical support internationally despite the fact that it did not command the will or power to bring its system into compliance or even have an interest in doing so.

30 The 46 countries that make up the Group of 33 advocate the maintenance of tariff protections for an array of Special Products and have pushed for the establishment of a Special Safeguard Mechanism as a means to protect their economies against import surges. The group is primarily concerned with creating an enabling environment for its members to scale up domestic food production

to meet domestic demand and thereby enhance food and livelihood security. Several G33 members criticized the July 2008 efforts of the WTO Director-General to use the global food price crisis as a rationale to lobby for a rapid conclusion to the Doha Round. They argued that the removal of distortions to agricultural trade alone would not foster sustainable production increases. In their view the food crisis will only be solved over the long run if a greater number of Special Products are created and a robust safeguard mechanism is implemented (*Inter Press Service* 2008).

31 Outside of the WTO the lack of a governed exchange rate regime is associated with similar development costs. In this unregulated context central banks in cotton-dependent countries have rapidly accumulated foreign exchange reserves. These can be drawn upon to protect currency values during times of excessive depreciation and can enable governments to avoid the harsh conditionality associated with the recourse to IMF lending. Foreign exchange controls and bond issuance facilitated this accumulation, and it had demonstrably high social and opportunity costs. Down to 2008 it resulted in an especially inefficient allocation of resources, diverting millions from investments in structural transformation to savings that were subject to low returns and foreign currency risk (Elhiraika & Ndikumana 2007). After the credit crisis struck reserves seemed inadequate to the new financial challenges. Also at the global level, donor resources deployed for the Millennium Development Goals (MDGs) have been insufficient, a fact that has led with few exceptions to an acute lack of progress on the goals across Africa (ECA 2005: 1).

32 Studies have also differed in their policy coverage, time horizons and base period. In addition to the conflicting parameters or assumptions detailed here, researchers have asked markedly different questions. See Sumner (2007) for a discussion of the divergent approaches.

33 The world lint price most often referenced is the 'A Index' price, a daily average of the five cheapest lint quotations maintained by Cotlook, an independent, non-trading company that publishes cotton news. Other important factors behind the 2008 price spike included increasing oil, shipping and insurance costs, though the extent and duration of the increase should not be overstated. According to the *Economic Report on Africa 2008*, cocoa, coffee, cotton and tropical log prices are more notable for the stability that they exhibited during the recent agricultural commodity boom (ECA-AU 2008: 2). By December 2008 the 'A Index' price had fallen back to 45 cents per pound, and subsequently hovered around the 50–55 cent mark into June 2009.

34 Even if a renewed focus on cotton ensured that producers could access the materials and training necessary to raise their output and reduced the time they spent on their cotton plots marginally, there is no guarantee that more attention to food crops or food production volumes would increase proportionately and in a manner that did not exploit women.

35 Food price rises are related to, *inter alia*: increased demand for grain and maize induced by scaled up livestock production to meet a structural shift in Asia whereby relatively richer consumers are adding more protein to their diets; the biofuel fiasco that has reduced the amount of land available for food crop production; weather and climate change-related production shortfalls; and the moves several key food exporting countries have made to export tariffs to boost domestic food security. Food price inflation abated somewhat as the

global recession took hold and oil prices dropped in late 2008, though the underlying structural trends remained intact.

36 These consumers are increasingly demanding products that have been weaved with fine cottons from Egypt and elsewhere. Egyptian cotton commands a premium of up to double the Cotlook 'A index' lint price. This market segmentation has been one of the factors that has lowered global demand for coarser machine-picked upland cottons, and could reduce future demand for handpicked Sub-Saharan cottons with short-staple lengths as consumers move up market.

37 One aspect of this evidence is that the unmanaged cotton price declined more slowly than the cocoa and coffee prices after efforts to manage the latter two prices were abandoned. Gibbon and Ponte attribute this phenomenon to the release of buffer stocks and also to the reemergence of secular price declines after supply management efforts ceased. Alfred Maizels (2003: 178) has nonetheless highlighted the positives associated with the more limited efforts to manage supply under the 1993 Cocoa Agreement. The International Cocoa Council now produces annual reports on the world market situation and six-year forecasts. With an eye towards balancing the market, producing countries can make use of this data to direct production levels towards meeting future demand trends predicted by consuming countries. This planning and market development mechanism for a product with a multi-year delay in the adjustment of production to shifts in demand is reflective of a minimalist approach to price management and market governance. Adherence is entirely voluntary in nature and the focus on price stability has not prevented significant price declines. The spectre of supply risk also induced a sharp price rise in 2008.

38 The UNCTAD Secretariat's annual flagship *Trade and Development Report* and *Least Developed Countries Report* continue to offer strong empirical analyses of the problems associated with commodity dependence. Beyond the ideological shift, higher food prices in the recent period could also exert downward pressure on the demand for governance arrangements to stabilize the prices of other agricultural raw materials at remunerative levels, despite the fact that the price rises of the latter have underperformed (Toye & Toye 2004: 234).

Chapter 4   Breaking the Historic Relationships in Tanzania

1 The labour issue is especially relevant for cotton farmers in the present conjuncture insofar as the spectre of acute rural labour scarcity looms ever larger. Environmental degradation and economic pressures have pushed people into off-farm income-generating activities while the pervasive influence of consumerism has pulled young people to seek opportunities for self-advancement in urban centres (Bryceson 1997). Though the extent to which income diversification has fostered 'deagrarianisation' is context-specific and contested (Yaro 2006), rising labour costs could undermine the ability of cotton to pay for those that grow it even if higher world prices are realized. Cotton remains labour-intensive *vis-à-vis* other annual and perennial (tree) crops. The introduction of productivity-enhancing techniques or technologies might only affect this status amongst well-connected farmers that are able

to gain access to these new resources if, when and where they become available.

2 Domestic governance reform and resource mobilization, development assistance, biotechnology policy, foreign direct investment and lint export dependence are assuredly not the only structural and policy factors that bear upon poverty outcomes. Structural food deficits and perpetual reliance upon imported inputs come to mind. Other policy priorities and policies are similarly noteworthy, such as (effectively depoliticized) inflation targeting or agriculture policy. However, I have not singled such topics out as they can be largely subsumed under the resource mobilization and governance heading. In my view, the dynamism and controversy surrounding biotechnology, FDI and the value addition imperative necessitated specific treatment. Policies that are implemented in each of these areas will have significant impacts on producer incomes and the capacity of cotton-dependent economies to reap better deals. Beyond aid, domestic governance reform and resource mobilization, these factors constitute the limits of the possible with respect to future earnings from cotton. That said, political and cultural factors operative in particular cotton-producing countries are of crucial import regarding the future distribution of opportunities from cotton. The ensuing discussion reflects this reality insofar as the governance/resource mobilization section is relatively large, and political and cultural considerations are peppered throughout the other sections.

3 Government figures indicate that rural poverty – in terms of the number of people living below the basic needs poverty line – has been 'dented' over recent years (Evans & Ngalwea 2003; England 2005). Regardless of the validity of this headline trend the majority of cotton farmers continue to lack capital and inputs, and toil for the 100 person days per season that it takes to grow one hectare of cotton barefoot with a jembe or hand hoe for low yields (Larsen 2006).

4 Credit needs could also be met if Savings and Credit Cooperative Societies (SACCOS) and credit guarantee schemes become more prominent in the Western Cotton Growing Area (WCGA). Mwanza has thus far progressed the farthest of the cotton-producing regions on these two fronts. There, Geita's District Commissioner has pushed hard for the formation of input-related SACCOS and DFID has backed the creation of at least 19 societies through Dunduliza, a non-governmental partner chaired by University of Dar es Salaam Professor Lucien Msambichaka. Several hundred farmers in Butobela, Geita contributed subscriptions of 5000 shillings each to an upstart SACCOS for input procurement during the 2007–08 marketing season. However, a few of the growers that I spoke with there were not able to pay the joining fee. This problem will have to be addressed moving forward. President Jakaya Kikwete's microfinance programme popularly known as 'JK Billions' has had notably little impact throughout the cotton zone. Constraints have included an urban bias in distribution and conditions that the designated lending agents – the National Microfinance Bank and the CRDB Bank – imposed on lending the government funds. Regarding credit guarantees, Professor Andrew Temu of Sokoine University chairs a trust known as Private Agricultural Sector Support (PASS) Tanzania that acts as a guarantor for outgrowers or other prospective investors. Temu has admitted that the guarantees PASS has issued in the past have been concentrated outside of the WCGA. He has expressed a desire to

remedy this underprovision problem. Enhanced microcredit and finance opportunities must also be managed in a transparent manner in order to ensure that higher producer debt levels do not generate perpetual indebtedness or degenerate into debt bondage. An exposé in *The New York Times* revealed the subjugation of cotton farmers in Tajikistan via the latter mechanism (Stern 2008).

5 Tanzania produces an American Upland type of cotton, *Gossypium hisutum L*, with a medium fibre or staple length that averages 27–28.5 millimetres. The Board maintains its own national standard system to measure the physical attributes of seed cotton that could affect lint quality or spinning efficiency. Of late, over 40 per cent of the crop typically has been graded 'strict middling GANY', where 'strict middling' denotes the second of four descriptive standards for GANY (GANY -1/4), a physical grade standard that indicates fair or average quality. A very small percentage has achieved the superior TANG grade, and over 50 per cent has been classed as the inferior YIKA grade. Roller ginned Tanzanian lint has a low content of tangled (nep), short or dead fibres, and the Board claims that over 95 per cent of production is in the prime range for the cross-sectional measurement of fibre maturity and fineness known as the micronaire. A high micronaire is a key determinant of successful carding, combing and dyeing (Worsham 2003). According to the Board the uniformity ratio of Tanzanian lint has also continued to be high and fibre strength remains in the medium range.

6 Tanzania enjoys a considerable cost advantage per hectare of rain-fed cotton *vis-à-vis* West African countries. The Board has estimated that in 2003–04 it cost the average farmer the equivalent of $177.50 to produce and market one hectare of seed cotton. If the figures were accurate, that producer would have only pocketed about $75 after covering their costs.

7 Improved access to health and veterinary services is certainly not a silver bullet. For example, malaria vector control through spraying household surfaces with DDT could improve farmer health over the near term but have a more ambiguous long-run effect. The introduction of affordable veterinary medicine could also enable farmers to strengthen their animals and improve the return on investments in draft power. However, if prescriptions of antibiotics are scaled up, farmers that subsequently consume these animals or the next generation of rural dwellers could be exposed to a risk that remedies for a range of tropical pathogens have been rendered less effective. The extended use of the low quality, cheap or pirated generic pharmaceuticals that make up as much as 30 per cent of the available supply is similarly hazardous (WIPO 2007).

8 The Board and the Tanzania Gatsby Trust are working towards a Cotton and Textile Development Programme to achieve a labour-intensive renewal of downstream industries. While the 12 operating spinning and weaving mills employ only 12,000 people, their rated capacity could meet up to two-thirds of domestic demand for textiles.

9 Colin Poulton (2006) has questioned whether yield estimates of 500 kilos per hectare are 'implausibly high' in certain districts.

10 W. Arthur Lewis (1965: 96) argued that productivity enhancements were a uniformly positive way to resolve distributional issues. In his view they could enable simultaneous increases in mass consumption and saving, and also boost tax revenues and therefore raise the level of expenditures that could

be made to achieve development targets in agriculture or other sectors. However, there is no guarantee that the traditional prescriptions to raise the productivity of cotton-producing households will generate pro-poor outcomes if they entail welfare-reducing trade-offs. When cash poor producers have to pay out of pocket to increase their own productivity or use materials that jeopardize their health, productivity enhancements are not a universal cure-all. If the Board's plan to expand cultivation extensively succeeds, many women cultivators might find themselves relying more on food imports and working similar hours on larger plots with more chemicals for a monetary reward that they do not control. A more nuanced view of productivity would help to square the sector-wide approach known as the Agricultural Sector Development Programme with the livelihood and benefit-sharing objectives found in Tanzania's Development Vision 2025.

11 Growth and poverty targets contained in MKUKUTA have been broken into three clusters: (i) growth and reduction of income poverty; (ii) improvement of quality of life and social wellbeing; and (iii) governance and accountability (URT 2005c). The CSP was largely silent on the Board's immediate actions or responsibilities regarding the latter two clusters save for the articulation of a deadline to mainstream 'cross-cutting' issues by 2010, and some language on the need to sensitize its employees to the governance issue. It did not touch upon the stunting and child labour problems that I witnessed in Mwanza, and failed to discuss how the static maternal mortality rate or the fate of HIV/AIDS orphans across the cotton zone could be better addressed.

12 Whether corruption is considered a process or an outcome related to the cultural economy of patron-client relations, or treated within a universalistic moral discourse or on a case-by-case basis, greater regulatory oversight and control of the vice remains a poverty reduction imperative (Khan 2006). Tanzania ranks 98th on Transparency International's Corruption Perceptions Index and over 88 per cent of all firms that responded to a 2006 survey on the topic reported paying a positive amount in bribes (Fjelstad et al. 2006; Lamsdorff 2007).

13 TCB operations today do not reflect all of the post-liberalization governance shortcomings Cooksey (2003) identified in his study of traditional export crop markets. The Board does not appear to have been 're-empowered', there is no evident 'hyper-taxation' and the aid regime no longer directly supports what he termed 'non-performing' administrative systems directly. The input, marketing, quality and poverty problems that have plagued the sub-sector suggest that a further factor Cooksey flagged – the persistence of 'ineffective' or unimplemented market regulation in these areas – continues to be a principal challenge.

14 Tanzania remains one of the poorest countries in terms of consumption per capita. Its poverty line was included in the calculation of the mean of 15 such lines that was recently used to revise the global absolute poverty line upwards to $1.25 per day (Ravallion & Chen 2008). Gross income per capita by some measures remains below $300. While the accuracy of statistics contained in the World Bank's *Africa Development Indicators 2007* is subject to the usual provisos regarding insufficient research capacity, data collection shortfalls and sampling errors resulting from the inaccessibility of information and communications technology, the figures are stark. Average life expectancy is 46 years,

69 per cent of the population is literate, 7 per cent have contracted HIV and only 2 per cent of the rural population has access to electricity.

15 Researchers who view aid in a more positive light have focused on the state of particular aid relationships and have also attempted to evaluate the development effectiveness of the Helleiner Report, its follow-ups and the impacts of the independent monitoring group that is building upon Helleiner's legacy. Mulley (2006) recently surveyed donor relations and questioned if Tanzania was indeed a genuine case of 'recipient' leadership. In 2005, the latter principle was labelled 'country ownership' and enshrined at the heart of the Paris Declaration on Aid Effectiveness (OECD 2005). This new approach to aid also entails greater donor coordination, harmonization and alignment with country systems, the streamlining of aid towards general budgetary support, and moves to advance the principles of mutual accountability and managing for results.

16 This shift was not captured in the 2008 Ibrahim Index of African Governance (2008). An initiative of the Mo Ibrahim Foundation, the Index attempts to measure governance performance across five categories: (i) safety and security; (ii) rule of law, transparency and corruption; (iii) participation and human rights; (iv) sustainable economic opportunity; and (v) human development. Tanzania slipped only one place to 15th of 48 on the 2008 list. It is possible that the equal weighting of categories and the reliance upon survey data and statistics from the 2005–06 period enabled Tanzania to avert a more substantial fall.

17 Several attendees at the Civil Society Organization (CSO) Parallel Forum on Aid Effectiveness held prior to Accra and other prominent development researchers based in the South have continued to challenge aspects of the Paris principles that they believe are conducive to donor interests (Reilly-King & Sneyd 2008).

18 The United States Agency for International Development (USAID) actively promoted biotech cotton containing the toxin-producing bacterium *Bacillus thuringiensis* (Bt) used to control pests over this period. In March 2007 USAID funded the attendance and participation of Cotton Board officials at a South African symposium intended to demonstrate the merits of Bt cotton. A national biotechnology policy had not yet been established.

19 Another type of GM cotton currently under development could have a greater impact on the drought-prone WCGA. A general cooperation agreement struck between the Israeli biotech company Evogene and the Centre de coopération internationale en recherche agronomique pour le développement (CIRAD) in June 2004 launched a programme of work that aims to develop a drought-resistant variety that will sustain yields even under conditions of extreme water scarcity.

20 These positions parroted opinions Monsanto and other leading players in the industry have articulated on the ability of Bt to raise yields, save labour time and improve producer health and the environment through reducing the usage of chemical inputs. Bt varieties are now planted on more than 28 per cent of the 35 million hectares of cotton cultivated annually worldwide.

21 CopCot Cotton Trading Ltd., a subsidiary of the international trading firm Paul Reinhart, conducted an experiment with input credit and credit-based input provision that failed after recipients sold their cotton to other traders

or failed to repay their loans. Backed by Remei AG, a Swiss-based organic yarn spinner, bioRe Tanzania's organic operation has maintained a durable outgrower-type arrangement. Copcot and bioRe's FDI-backed works are discussed in Chapter 6.

22 Several interviewees claimed that the Government also has evidently deficient capacities or incentives to enforce aspects of the Environmental Management Act 2004 that could foster pro-poor rural outcomes over the long term, or the provisions of the Labour Act 1997 related to hours of work and child labour. In the absence of a mandatory global code of conduct for FDI agricultural investors have few reasons to ensure that their operations are in full compliance with laws that might cut into their bottom lines. Investors only need to hear out the TIC's informal requests for skills and technology transfers and for local empowerment or equity stakes. There are no formal mechanisms in place to guarantee that the few investors actually interested in agriculture will take up TIC calls. The TIC is a self-described 'facilitator', not an investment cop.

23 In spite of the US African Growth and Opportunity Act and the fact that at least one prominent report (UNCTAD-ICC 2005) has characterized the textile sector as a promising investment 'opportunity', high power tariffs, frequent outages and the offer of comparatively weak incentives have dissuaded many downstream investors from choosing Tanzania.

24 Gibbon and Ponte (2005: 137) claim that the only performance requirement lint suppliers have faced is timely delivery of the baled lint to a port. International traders have also only upheld and arbitrated conventions governing cotton quality. As such they have apparently had a relatively lighter touch than other commodity merchants who have dealt with African suppliers.

25 The United Nations Industrial Development Organization (UNIDO 2005) has helped several West African ginners in cotton-dependent countries to procure the high-volume instrument.

26 The 2004 Participatory Poverty Assessment (PPA) concluded that HIV/AIDS was arguably the single most severe factor of impoverishment in rural Tanzania (URT 2004b: 122). Many HIV-positive people in the countryside do not simply turn to Western medicine, and resort instead to the social custom of witchcraft. While respondents upheld the view that HIV/AIDS was a scourge, the report's authors disagreed with the local belief systems and norms that maintained the power of witchdoctors and structured the thinking of poor people on how to 'solve' this health challenge. The PPA characterized witchdoctors as a barrier to poverty eradication, though the relationship of this conclusion to the opinions people expressed was not clarified.

Chapter 5   NGOs, Conventional Production and Poverty

1 NGOs can be defined broadly as private, voluntary or non-profit groups whose primary aim is to influence some form of social change (Khagram et al. 2002). In Scholte's view, these actors can educate, give voice, fuel debate, increase transparency or promote more effective governance. However, he also argues that they can pursue dubious goals, promote flawed or

poorly conceived policy, detract from the idea of governance itself or be coopted by governance institutions.

2 Those that view the Initiative in a uniformly positive manner seem not to have stopped to consider the history of the man who was chosen to address WTO delegates. Compaoré came to power in a bloody *coup d'état* that deposed President Thomas Sankara in 1987. By all accounts Sankara had dedicated his Presidency to advancing the principle of poverty eradication, albeit from a Marxist point of view. After Sankara's murder Compaoré repealed all of the reforms that Sankara had implemented to liberate the countryside from exploitative neo-colonial relationships, traditional authority structures and corrupt practices. See *Thomas Sankara: The upright man*, a 2006 documentary film by UK filmmaker Robin Shuffield.

3 With the sole exception of the powerful US and European cotton lobbies, lint producers, international cotton traders, yarn spinners and other nominally private participants in the global cotton value chain were not initially engaged in this transnational network. While most definitions of the network phenomenon typically exclude actors that do not share the 'common values' espoused by network participants, research conducted by William D. Coleman and Sarah Wayland (2006) suggests that a failure to consider these dissenters can unduly limit analyses. To ignore this segment of oppositional network participants would rule out attention to the groups that encouraged several US senators to defend cotton subsidies. These entities also worked as a protectionist coalition in the lead up to the 2003 WTO Ministerial at Cancún to prevent the US Trade Representative from offering concessions that would undermine the interests of US cotton.

4 National cotton growers' unions came together at Cotonou on 21 to 22 December 2004 to form the Association des Producteurs de Coton Africains (APROCA). Like ROPPA, this organization can be characterized as an emerging transnational social movement. Transborder solidarity is evident and it is likely capable of coordinating sustained political mobilizations across borders.

5 Sikkink (2002) has argued that the NGOs with the most resources at their disposal are most likely to have access to and wield influence at the international level.

6 Sadly, the final point is an important one. These efforts have been largely if not exclusively delimited to the West, a fact that the otherwise rigorous evaluations of this campaign failed to raise or problematise.

7 IDEAS had commenced exploratory research on West African trade concerns for the Swiss the previous September, and this initial project wrapped in April 2003.

8 One such instance was the Centre's push to have President M. Amadou Toumani Touré of Mali present the Initiative to the Trade Negotiations Committee and thereby improve its optics. Blaise Compaoré was convinced to go to Geneva only after efforts to bring Touré failed. A discussion of the submission and the ways supportive non-governmental actions could be coordinated cohosted by IDEAS and ICTSD on 6 June 2003 also underscored the political nature of the Centre's work.

9 Despite this inauspicious start, the paper published several follow-up pieces on the road to Cancún, including one that was directly attributed to an IDEAS analyst.

10 This message was apparently lost on the President and did not resonate in the US cotton belt. Republican Senators Chambliss, Cochran and Lincoln drafted a letter on 8 September to then US Trade Representative Robert Zoellick to express their displeasure with the Initiative. They portrayed it to be in fundamental opposition to the interests of the United States and encouraged Zoellick to pay no heed to the African position. When Zoellick penned an open letter to trade ministers four months after Cancún on 11 January he nonetheless indicated a willingness to move on cotton. A subsequent meeting between IDEAS staff and two members of the US mission to the WTO confirmed that the USTR viewed the Centre as a 'valid and interesting' partner. This limited impact – the USTR's explicit recognition of a cotton problem and its conditional endorsement of liberalization – can be characterized as the outgrowth of IDEAS work.

11 Hazard was particularly concerned about the lack of opportunities cotton growers had to participate in domestic and regional governance arrangements that had impacts on their livelihoods. He argued that a more appropriate agenda would have been to fight against 'political poverty' within Africa, and that strategies to exit the 'crises' associated with cotton would have looked much different if farmers had been able to exercise voice.

12 Opponents of the Sectoral Initiative might now argue that the Centre's work actually did more harm to the negotiations and even the system itself. While the Multi-donor-C4-IDEAS project can be described as an attempt to foster the socialization of African governments to trade principles, critics charged that the Initiative disregarded the 'box' approach, an important unifying principle of the agricultural negotiations. They also argued that the compensation request was not rooted in WTO law. In this light the Centre has politicized the WTO and raised expectations that if left unfulfilled could undermine the future of the organization.

13 The external evaluation of the project conducted during the fourth phase found that 'partner' countries had raised questions about who exactly exercised 'ownership' over these interns. Several officials appreciated the tactics and proposals that had emerged from the technical antenna, but complained about the direction of the decisionmaking process. Even though the C4's Geneva-based representation had increased significantly since 2002, the antenna bypassed Geneva and transferred its propositions directly to the national level, a practice that continued down to the conclusion of the third phase in December 2006.

14 Non-governmental sources of development finance could be classified as the 'growth industry' in the political economy of aid. According to statistics compiled by the OECD-DAC and the latest available figure on annual UN system-wide expenditures, the grants international NGOs and foundations provide to their affiliates or 'partners' in the South each year are roughly equal to or even greater than the costs of maintaining the UN system (Global Policy Forum 2006). In 2006, a record $14.8 billion in 'net private grants' flowed to developing countries while official donors provided $2.4 billion to NGOs south of the Sahara.

15 One of my interviewees characterized this work as 'strictly ideological' insofar as farmers were mobilized on the grounds that 'global forces' or factors were 'ripping them off'. This individual was highly critical of the failure of international NGOs more generally to inform cash and food crop sellers that the

shady dealings of private sector players and officials from the village, ward, district, board, and ministry levels have been a perpetual source of hardship.

16 In the heart of the cotton zone only several non-governmental and foundation partnerships with community-based organizations have been established. Oxfam Great Britain, for example, has supported the Youth Advisory and Development Council (YADEC), and the bioRe Foundation has worked with Wadec, a women's group. The initiatives that churches and faith-based organizations such as CARITAS, the Catholic relief, development and social service organization, have undertaken in the region to alleviate suffering should also not be downplayed.

17 Amongst city dwelling policy elites there was also an impression that several for-profit consultancies had sprung into existence. These interviewees claimed that after more money became available for HIV/AIDS outreach and research such for-profit entities seemed more interested in accumulating income-generating projects than in delivering actual results.

18 The need for greater coordination of NGO activities was raised by nongovernmental activists at the third high-level forum on aid effectiveness held at Accra in September 2008. The Accra Agenda for Action (OECD 2008) welcomed the proposal these actors tabled to initiate a multistakeholder dialogue in order to promote the development effectiveness of their own operations. If realized, this process would aim not only to improve intra-NGO and NGO-government coordination, but also enhance NGO accountability and scale-up knowledge of their activities in the private sector. The latter problem is especially evident in Tanzania where as Professor Sam Wangwe put it and other interviewees confirmed, most private enterprises simply 'do not know what NGOs are doing'. Wangwe would like NGOs to be integrated into district and regional planning processes. In his view the realization of greater complementarity would reduce the possibility that investments will be distorted or skewed in the future due to a lack of engagement with district officials or alignment with their plans.

19 Extension agents at the ward level were relied upon to implement the project and are now responsible for keeping tabs on the agro-pastoralists that received a recovery package. Many of these people were conventional or organic cotton farmers. Oxfam's confidence in extension agents might be somewhat misguided given that many people throughout the WCGA told me that extensionists rarely visit cotton farms.

20 If goat giving resulted in fewer cattle coming to the market during crisis-times cattle prices would not drop as low as they had in the past, when sellers received only 20–25 per cent of the pre-crisis value at the market. Risks of this approach included the generation of local inflation, the unknown prospects for the future secondary market in goats and the possibility that there could be negative cultural impacts or social stigmas associated with goat ownership. The latter two topics are worthy of further anthropological study.

Chapter 6   CSR and the Cotton-Poverty Relationship

1 Bono's advocacy has encouraged consumers to focus on only one aspect of extreme poverty, and discouraged discussion of the factors of impoverishment

operative in the RED™ value chain itself. Beyond this problematic commodity fetishism, Richey and Ponte have claimed that the reliance on celebrity RED™ exudes is contradictory to a hardcore certification and audit culture, though they are reticent to dismiss it entirely as it has raised tens of millions for the Global Fund.

2 Even Michael Conroy (2007), a prominent CSR advocate, admits that there are limits to the business case for responsibility. Corporations typically evaluate the prospective benefits of adherence to higher standards, including lower insurance and financing costs, and reduced reputational risks, and weigh these against the short-term costs of compliance. The fact that equity markets continue to reward or punish companies primarily for financial rather than social performance (Vogel 2005) has led Blowfield (2007) to suggest that amongst publicly traded firms the correlation between doing well in financial terms and doing good for society is weak at best.

3 These governance institutions derive their authority directly from interested audiences and not from sovereign states. According to Bernstein and Cashore (2007), NSMD governance seeks to embed social and environmental norms in the global marketplace.

4 It is likely that many Southern-based fair trade activists would be nonplussed to learn that their budding alternative had been one-sidedly designated as a 'market driven' form of non-state governance given their aspirations for a bottom-up, producer-driven supply chain that seeks to serve and expand new markets while not being subservient to the market logic of price competitiveness.

5 Adidas, Gap Inc., H&M, IKEA, Conservation International, the Interchurch Organization for Development Cooperation, Organic Exchange, Oxfam, the United Nations Environment Programme and the WWF were members of the BCI steering committee at the outset. Several of these organizations, the Swedish International Development Agency and the State Secretariat for Economic Affairs of the Swiss Confederation funded the creation of a project support team, and the WWF executed this task. Other enterprises and groups played an active role in the BCI on many issues and contributed financial or in-kind support, including Cotton Made in Africa, KappAhl, Lindex and Marks & Spencer, and these came to be known as 'BCI partners'. The steering committee also supported the creation of an advisory committee composed of 20 individual experts to provide technical advice during the consultation and pilot phases.

6 One steering committee member has promised to source better cotton exclusively from 2015.

7 Other publicly available BCI reports used this exact terminology. See for example BCI (2007).

8 Stakeholders that attended this workshop either hailed from or were based in Bénin, Burkina Faso, Cameroon, Mali, Sénégal and Togo. Other cotton-producing countries of the sub-region, including Chad and Côte d'Ivoire, were not represented.

9 Contributors to the discussion further revealed the limited nature of the project when they raised questions about transboundary issues, the implementation of existing national regulations or other poverty-reducing policy options that the project support team did not have the power to address.

This private, voluntary initiative simply could not target all matters that came up in discussion, including the cross-border migration of livestock or low-cost farm labourers, and the evident disjuncture between regulatory texts and on the ground enforcement. Several participants also expressed the idea that the BCI could be fashioned into a vehicle to stimulate the obligatory uptake of welfare-enhancing technologies such as nitrogen-fixing leguminous plants, and the reality of its limited authority similarly frustrated their ambitions.

10 This meeting demonstrated just who would exercise 'ownership' over the better cotton process and which 'Africans' it would be accountable to. Concerned parties from major cotton-producing zones in the west, east and south of the continent beyond the trial area were once again excluded. Later in 2008 the support team skirted this issue when it boasted that 32 per cent of all stakeholders who had been consulted were African nationals (BCI 2008c). The marginalized could apparently find solace in the team's faith that outsiders would reap future benefits from the learning by doing that the early movers would engage in, and the sharing of 'tested best practice' that would ensue.

11 West and Central Africans that were consulted voiced a concern that the Initiative did not spell out the specific benefits farmers could expect (BCI 2008b). They also questioned the lack of tangible objectives for gender and intra-household divisions of labour and resources, and wondered about the issues of food security and land rights. On the latter, they were told that the steering committee needed 'more time' to consider how these issues related to 'better' cotton.

12 A paradox of the involvement of socially and ethically concerned NGOs in the BCI process is that they might now be complicit in generating an unforeseen threat to the fair trade model. Down to 2005 corporate codes had posed negligible challenges to the viability of fair trade certification and markets due to the fact that they were not often promoted to consumers directly (Nicholls & Opal 2005: 141). Members of the BCI now have the ability to do just that at their own discretion. At worse, branded retailers in the scheme might take advantage of this freedom to produce advertisements and marketing campaigns that conflate better cotton with 'fair trade'. By early 2009 the BCI had not discussed or designed mechanisms to restrain the development of misleading propaganda that could do serious harm to the fair trade movement.

13 CMIA partners include Accenture, Avery Dennison, the German environmental association NABU, the German Federal Ministry for Economic Cooperation and Development (BMZ), the German Investment and Development Company (DEG), the German Society for Technical Cooperation (GTZ), Dunavant SA, McCann Erickson, Otto Group, Tchibo gmbH, Tom Tailor, Welthungerhilfe and the WWF. A group of allied manufacturers and retailers comprised of 1888 Mills, Anson's Herrenhaus KG, Bierbaum Unternehmensgruppe, Frankenstolz, Otto Group, QVC and Tchibo GmbH has also been formed. Alterra, Faso Cotton, Christoph Leuschner, Stiftung Umwelt, and Paul Reinhart AG are the principal sponsors.

14 From the initial missions to kick off the pilots in 2006 the project proceeded in three stages. During the first phase local partners conducted self-assessments of their adherence to the new standards detailed below. The second phase introduced an independent verification system, and a broader rollout was planned for the third phase.

15 Promotional materials were not entirely devoid of realism. The relatively high transactions costs historically associated with storage, transportation and forwarding from Africa were flagged as considerable threats to the project's viability. Nonetheless, the Foundation did not disseminate any information that could be taken to demonstrate their knowledge of the local cultures of social and economic advancement that maintained these costs, or any plan to work with people in a culturally sensitive manner to ascertain, what, if anything, could be done to remedy this issue.

16 Not all prescriptions are as outwardly progressive. The conspicuous politics of the standard for competition is case in point. Green thresholds will only be reached on this front once at least 75 per cent of eligible smallholders are able to obtain pre-financing from CMIA-affiliated buyers in a competitive buying environment. As such, this provision could potentially impact the development of CMIA certification in countries where single-buyer, monopsony systems continue to function. A failure to thoroughly analyse the politics or political implications of the standards is also evident in the prescription regarding the necessity of written contracts for input supply and sales. Literacy rates in Bénin and Burkina Faso have recently stood at 35 and 22 per cent respectively. Absent an objective regarding the education of adult farmers, the provision of resources to such ends and efforts to build the capacity of governments to ensure that weaker counterparties in remote cotton-producing zones have access to subsidized legal advice, the arbitrary imposition of written contracts could generate a range of impoverishing outcomes, including higher incidences of debt bondage and fraud.

17 Dunavant has claimed that its field schools on integrated pest management, extension staff training and demonstration plots on the CMIA pilot in Zambia have improved the abilities of its contract farmers to prepare their lands, time their planting and protect their crops.

18 To respond to offences, activists could target individual firms or the 'certification' system itself. In my view, the latter point of departure would test CMIA's responsiveness to civil society demands, and also its capacity to maintain and attract the interest of market participants. This approach, like other private standards initiatives (Tallontire 2007), was created to reduce firm level risks and leverage the new market for virtue. However, its current form invites the risk that it will be deemed to be an elaborate façade.

19 Like the Otto Group, an independent third-party certification agent has verified Remei AG's compliance with Social Accountability International's SA 8000 global social accountability standard. Adherence to SA 8000 indicates a commitment to proscribe child labour, forced labour, and discrimination, and to ensure that other universal human and labour rights are respected, such as freedom of association, collective bargaining and fair work hours and remuneration. In my view, a third-party assurance that no child labour has been used in production is especially significant as regards cotton production in Tanzania. During my field research I witnessed a child labour team first-hand as they weeded a large conventional cotton plot on a school day.

20 The latter insight was just one of several examples of learning by doing and continuous improvement that I observed directly on my visit. According to staff, many field officers had simply visited farms to collect data and failed to take advantage of opportunities to advise and chat with farmers during

the project's formative years. In 2007 they expressed a desire for greater mobility and social interaction that if acted upon, could buttress the enabling environment for poverty reduction evident in Meatu. Similarly, it took some time to arrive at a culturally appropriate approach to engaging with the third-party certifier. For years the Institute for Marketecology (IMO) had flagged management for the inconsistency with which it acted upon the certification agent's recommendations. After some time it became apparent that cultural inattentiveness was at the root of this shortcoming. Staff were subsequently sensitized and coached on the need to respond directly and in an upbeat manner to all questions posed by the certifier – even if they had been framed in ways that would normally be interpreted as aggressive or worse – and communication improvements ensued that made possible greater levels of conformity. The willingness and demonstrated ability of this operator to learn and improve stands in marked contrast with standard buyer behaviour in Tanzania's conventional cotton market.

21 bioRe paid 10 TZS per kilo of seed cotton that it purchased into the CDF. Conventional buyers whose 'suppliers' actually made use of these inputs paid 20 shillings. During the 2006–07 marketing season bioRe purchased 4.581 tonnes of seed cotton and paid the USD equivalent of 45.81 million TZS into the fund (approximately $36,346 at the March 2007 exchange rate).

22 Remei AG afforded bioRe access to a level of working capital that would have come at a punishing cost through local financial intermediaries or been simply unavailable during its start-up phase. The parent also absorbed bioRe's early operating losses. These subsidies enabled investments in human capital and learning that eventually helped the project to turn a profit. New investors that choose to back copycat operations will have to take a similarly long-term view while transactions costs remain high and negative externalities from the conventional cotton system persist. Human capital and knowledge spillovers from the first-mover do, however, have the potential to reduce the amount of financial support and length of time that it will take for new entrants with viable market outlets to become self-sustaining.

23 The latter task could be a complicated one in other districts of Mwanza, Shinyanga, Singida, or Tabora regions where average farm sizes are much smaller than the norm in Meatu. bioRe has historically contracted farmers whose arable holdings have totalled no less than 3.6 hectares. Its production manager has contended that the three-year crop rotation system necessitates at least that much land to be ecologically and economically viable. Expansion to areas where smallholdings average less than one hectare would therefore require a substantially revised approach to crop rotation and possibly the participatory creation and implementation of a plan to boost cultivation extensively that has been vetted for sustainability and equity.

24 Grolink, a Swedish firm, provided organic certification and marketing advice for EPOPA.

25 In 2004 Haider had partnered with a non-governmental organization to produce organic cotton in Tabora. This project failed to get off the ground due to coordination failures and the suspicions of farmers that they would not benefit from it. Haider has admitted that it can take some time for people 'to believe that you are there to help them' because 'they have been left alone for so long'. In Singida rural, a district that he considers even

more promising than Meatu due to its climate and the explicit support of the District Commissioner, he is endeavouring to prove that organics can simultaneously enable input costs to shrink and yields to increase.

26 First hired by Tansales in 1995 while still employed at the Ministry of Agriculture, Kapanda is an organic cotton pioneer. He has claimed that many of his productivity-enhancing tips can be found in the most comprehensive manual on the topic, though the European-based authors did not attribute any of these ideas to him directly (Eyhorn et al. 2005). In Kapanda's view, '90 per cent' of the available material on effective production techniques and the agronomy of organic cotton presented in the guide and elsewhere is 'stuff learned from me'.

27 According to Organic Exchange (2007, 2008), organic cotton was cultivated on over 161,000 hectares in 2008 and 0.55 per cent of global lint production or 145,872 tonnes were ginned to organic standards. Wal-Mart, Nike, Woolworth South Africa, Coop Switzerland and others drove demand for this lint 118 per cent higher per year between 2004 and 2006. Lint buyers paid $66 million for organic product during the latter year while retail sales of products derived from organic cotton grew to $1100 million, a figure that was projected to treble by 2008. The evident disjuncture between the 2006 lint and retail sales figures indicates the depth of the value addition challenge. If African organic producers do not move to add more value they will continue to rely upon lint sales that constitute only 3 to 6 per cent of the final retail prices of organic garments, and if present trends hold, depend increasingly on Wal-Mart. This firm is notorious for squeezing supplier margins, and over-reliance on it could lead exporters into an intractable ethical quagmire.

28 Article 2.7 of the WTO Technical Barriers to Trade Agreement established the equivalence of technical regulations – even if these regulations differ – provided members are satisfied that these adequately fulfil the objectives of their own regulations (Coleman & Reed 2006). The result of a multistakeholder process supported by UNCTAD and UNEP, the EAOS aims to harmonize regulations governing organics within the regional market, and to serve as a basis for negotiations that will enable organic products exported from East Africa to be treated and labelled as such in Europe, the United States and Japan. It is based on standards in place across the region, the basic standards of the International Federation of Organic Agriculture Movements (IFOAM), and the Codex Alimentarius guidelines for organic production, processing, labelling and marketing.

29 Several interviewees involved with the TOAM noted that progress on certification and in particular the accreditation of a domestic certifier was a crucial imperative. The EPOPA programme had previously helped to establish the domestic certification body TanCert and had also supported the development of Tanzania's organic standards. As EPOPA wound down, TanCert applied to the International Organic Accreditation Service in January 2007 for accreditation against the IFOAM rules. Thus, the organics movement was able to argue that there was an urgent need to find a substitute source of support. If successfully accredited, TanCert would be able to issue organic certificates for the export market and as a consequence, cut certification costs and preserve foreign exchange that had previously leaked abroad to the IMO.

30 For example, one prominent economist argued that the broader challenge of poverty reduction could be better addressed through scaling up the domestic supply of staple foods for middle-income and elite consumers. He also warned that reliance on elite consumption patterns abroad necessitates a risky bet that any new sanitary and phytosanitary measures (SPS) imposed in those markets to cover new categories of 'special' products will not simply be thinly veiled protectionist attempts to support domestic producers and exclude African produce.

31 This research concluded that producer groups were more transparent and that the inputs women made were taken more seriously and acted upon post-certification. Even so, it found rampant absenteeism and the unchecked low-level of literacy to have impeded community empowerment.

32 According to other interviewees, conventional buyers had 'contributed' so little to productivity enhancement that it was difficult to make the case that any had 'fully engaged in helping farmers'. When I asked producers themselves about private service provision on the farm, one lifelong producer explained to me that company representatives no longer visited him to offer advice, and that these people now came to his ward only to determine the best buying locations.

33 One ginner contended that his outsourcing approach was indicative of his full participation in 'poverty alleviation'. His operation advanced its buying agents the cash necessary to make purchases and paid a 15 TZS per kilo fee for services rendered. He preferred to work through the old primary society structures that were still in place and able to organize buying, and claimed that he had advised society leaders to spend their money on 'community development'. Nonetheless, he admitted that just how the commission is split up is 'really their business', and hoped that primary societies would be able to execute buying and contract transporters without having to rely on cash advances in the future. In his view, advances were a risky business as many societies had contracts with several buyers and their leaders could be bribed to direct purchases towards a particular company. When society leaders were bribed in this way the working capital of non-bribe paying operations that advanced funds was effectively misallocated.

34 A few respondents wondered if there were any individuals at the board who had taken payments to bury these cases, but also expressed concern regarding the capacity of dirty ginners to adhere to minor regulations, embrace legitimate accounting practices or learn about the industry itself. Regarding the former, the Cotton Act stipulates that buyers must own their own storage spaces, and it had come to light that unscrupulous trader-ginners or agents had rented houses with mud floors to store their purchases, a cost-cutting measure that reduced product quality. These interviewees also speculated that perhaps only three of 28 firms that purchased cotton in 2006–07 made use of credible or reliable accounting firms. One even suggested that traders want 'to make money, but many of them do not know anything about the micronaire or the staple length…. All they know is that if they have a ginnery they can produce lint'.

35 Instead of squeezing the maximum amount of work from a limited number of employees, for instance, it hired on relatively high numbers of seasonal employees each year, including cleaners, canteen workers, off-loaders, loaders,

sorters, stackers, roller builders, foremen, press operators, press attendants, gin stand cleaners, seed stackers, gin feeders and godown (warehouse) feeders.

36   In March 2007 over 242,000 kilos of seed cotton Copcot's agents claimed to have purchased the previous season was unaccounted for. At the high farm gate prices that prevailed during the drought year the full deficit had the potential to cost the firm 80–100 million TZS, though employees believed that at least some of the missing cotton would be found.

37   Cooksey also critiqued Oxfam's trade campaign. He believed that efforts to make trade fair had furnished the Tanzanian 'political-bureaucratic elite' with rhetorical ammunition they could draw upon to rationalize heavy-handed interventions in 'liberalized' markets. In my view his position on this matter was akin to shooting the messenger, but was understandable given the emphasis he placed on the domestic level of analysis and Oxfam's near exclusive focus on global factors.

38   This decline stemmed from a sharp reduction in Chinese and Southeast Asian lint imports. Manufacturers there had excess lint supplies as orders for textiles and clothing from US branded retailers had weakened.

39   Interestingly, this report referenced research that focused only on the origins of the Meatu cotton project and its early years and did include any findings from research that had been conducted since its conversion from Tansales to bioRe Tanzania. In 1994 the Tanzanian cotton firm CIC approached GTZ, the German development cooperation group, to conduct a feasibility study on organic cotton. Producers were organized in Meatu later that year and Tansales was born, a project the UNEP-UNCTAD study praised for increasing the amount of food locally available and enhancing natural, social, human, physical and financial capital.

Chapter 7   Conclusions: Global Interventions and Poverty Eradication

1   In this issue area, I endorse Paul Feyerabend's (1975) call to let a thousand theories bloom with only two provisos. It is vital that the evident profusion of new research on this topic continues to struggle towards elusive truths and does not degenerate into elitist dialogues or monologues that are unintelligible or inapplicable beyond the ivory tower. As more stories continue to be produced and disseminated it is also important for researchers to consider their possible contribution to the state of knowledge and ask themselves whether resources would be unduly diverted from other equally worthy topics if they were to add their voice to this debate. Academic freedom in this purported 'new era of responsibility' is assuredly a principle to be advanced, respected and defended. I do, however, believe that social 'scientists' need to be attuned to the broader social implications of their research programmes, and hope that more of them choose to exercise their freedom while being evermore aware of the knock-on effects of their choices and actions.

2   Now that official donors and non-governmental organizations are moving to harmonize and align their assistance provision strategies and programmes with the stated priorities of recipient or 'partner' governments, country-level coordination is becoming more complex. Outsourced service providers and

farmer advocates will rely increasingly on the state itself to disburse the funds necessary for their own survival. The bleak prospect moving forward is that organizations that push the envelope a little too much could be starved of funds, an eventuality that could exert downward pressure on anti-poverty objectives and result in even more piecemeal progress.

Appendix B   Stabilization, Adjustment and Rural Livelihoods

1 Detractors included the small size of domestic markets, burdensome regulations, the fear of expropriation and nationalization, and outright statutory bans. Sir Hans Singer (1950) and Ragnar Nurkse (1953: 91) had previously articulated the political and economic rationales for restrictive approaches to FDI. Citing the example of Japan's industrialization Nurkse had argued that government investments that relied upon foreign financing could establish a more 'balanced development' of public infrastructure and services – what he referred to as 'social overhead capital' – than reliance on FDI alone. In his view export-oriented FDI was inherently 'lop-sided' or prone to geographic concentration in extractive zones. He argued that most citizens would reap benefits from it only after a sustained public investment programme.

2 High nominal exchange rates also worked against export competitiveness (Helleiner 1999).

3 The impact of cost-price spirals on the continent during this time was especially deleterious given the prevalence of the tendency for international inequalities in consumption opportunities to generate low savings levels amongst elites that were exposed to Western consumption norms and ostentatious displays of wealth (Nurkse 1953). Moneyed Africans that could spend their incomes legitimately or otherwise rationally chose to do so, and their habit of looking abroad to make final consumption choices negatively impacted balance of payments situations that were already shaky due to terms of trade declines (Wangwe & Semboja 2002).

4 Gibbon et al. (1993) emphasized this point in their extensive study of the Bank's involvement in African agriculture. Advocates of agricultural adjustment asserted that it would be a cost-effective means to promote specialization, output growth and higher incomes. Faith in the idea that it was possible for the market to get prices 'right' in Africa held despite Galbraith's (1967: 39) earlier insight that corporations in the most highly industrialized countries often made deliberate efforts to get prices 'wrong' through planning and controlling their input and output markets. Critics of adjustment later claimed that a sole focus on boosting incomes could not benefit the poor. In their view, equal attention had to be paid to protecting the productive assets that the poor commanded or had rights to, and preserving and bettering the access that they enjoyed to education, employment opportunities, extension services, healthcare and in-kind supplemental resources (Cornia et al. 1987; Woods 2006).

Bibliography

AfDB (African Development Bank) (2008) 'Interview with AfDB Chief Economist, Louis Kasekende: Escalating Oil Prices Pose a Challenge to Many African Economies', *African Development Bank: Press Story*, 14 March 2008. http://www.afdb. org (Accessed 26 March 2008).

Agazzi, I. (2010) 'Whither African Cotton Producers After Brazil's Success?', *Inter Press Service*, 23 March. http://ipsnews.net (Accessed 1 April 2010).

AGRA (Alliance for a Green Revolution in Africa) (2007) 'Revitalising African Agriculture: The Alliance in Action'. http://www.agra-alliance.org (Accessed 1 September 2007).

Akyüz, Y. (2005) *The WTO Negotiations on Industrial Tariffs: What is at Stake for Developing Countries?* (Geneva: Third World Network).

Amehou, S. (2005) 'Consequences and Challenges of the July Framework Agreement for the Cotton Case', in E. Hazard (ed.) *International Trade Negotiations and Poverty Reduction: The White Paper on Cotton* (Dakar: Enda Diapol), 23–32.

Anderson, K. and Martin, W. (2005) 'Agricultural Trade Reform and the Doha Development Agenda', *The World Economy*, 28: 9, 1301–1327.

Anderson, K. and Valenzuela, E. (2006) *The WTO's Doha Cotton Initiative: A Tale of Two Issues*, Policy Research Working Paper WPS 3918 (Washington, DC: World Bank).

APROCA/HEC/FARM (2007) Launch of the Cotton University. http://www.farm-foundation.org (Accessed 20 October 2008).

Appadurai, A. (2000) 'Grassroots Globalization and the Research Imagination', *Public Culture*, 12: 1, 1–19.

Atkinson, J. and Scurrah, M. (2009) *Globalizing Social Justice: The Role of Non-Government Organizations in Bringing About Social Change* (Basingstoke, Houndmills: Palgrave Macmillan).

Auld, G., Bernstein, S. and Cashore, B. (2008) 'The New Corporate Social Responsibility', *Annual Review of Environment and Resources*, 33: 413–435.

Baden, S. (2004) '"White Gold" Turns to Dust: Which Way Forward for Cotton in West Africa?', Oxfam Briefing Paper 58 (Oxford: Oxfam International).

Baden, S. (2005) 'Between a Rock and a Hard Place: The Case for Support to Africa's Cotton Sectors', in E. Hazard (ed.) *International Trade Negotiations and Poverty Reduction: The White Paper on Cotton* (Dakar: Enda Diapol), 97–108.

Baecker, G. (2004) 'Relevance of Risk Management Instruments for the Cotton Sector in West and Central Africa', in E. Hazard (ed.) *International Trade Negotiations and Poverty Reduction: The White Paper on Cotton* (Dakar: Enda Diapol), 125.

Baffes, J. (2002) *Tanzania's Cotton Sector: Constraints and Challenges in a Global Environment*, Africa Region Working Paper Series No. 42 (Washington, DC: World Bank).

Baffes, J. (2004a) *Cotton: Market Setting, Trade Policies and Issues*, Policy Research Working Paper WPS 3218 (Washington, DC: World Bank).

Baffes, J. (2004b) 'Tanzania's Cotton Sector: Reforms, Constraints and Challenges', *Development Policy Review*, 22: 1, 75–96.

Baffes, J. (2007) 'The Cotton Problem in West And Central Africa: The Case for Domestic Reforms', *Economic Development Bulletin*, Issue 11, 10 July 2007 (Washington, DC: The Cato Institute).

Baillaud, E. (1903) 'Cultivation of Cotton in Western Africa', *Journal of the Royal African Society*, 2: 6, 132–148.

Bassett, T.J. (2001) *The Peasant Cotton Revolution in West Africa: Cote d'Ivoire 1880–1995*, (Cambridge: Cambridge University Press).

Bates, R.H. (2005a) *States and Markets in Tropical Africa*, 2nd edn (Berkeley: University of California Press).

Bates, R.H. (2005b) *Beyond the Miracle of the Market: The Political Economy of Agrarian Development in Kenya*, new edn (Cambridge: Cambridge University Press).

BCI (Better Cotton Initiative) (2006) *Research Report on Global Social and Labour Issues in Global Cotton Cultivation*. http://www.bettercotton.org (Accessed 1 December 2008).

BCI (2007) *West & Central Africa Working Group Meeting 1: Summary Report*. http://www.bettercotton.org (Accessed 1 December 2008).

BCI (2008a) *Global Principles, Criteria, Enabling Mechanisms: Version 1.0*. http://www.bettercotton.org (Accessed 1 December 2008).

BCI](2008b) *Second Regional Working Group Meeting for West & Central Africa: Summary Report*. http://www.bettercotton.org (Accessed 1 December 2008).

BCI (2008c) *BCI Consultation Report: Global Principles, Criteria, Enabling Mechanisms*. http://www.bettercotton.org (Accessed 1 December 2008).

Beattie, A. (2008) 'Governments Can No Longer Ignore the Hungry', *Financial Times*, 4 April, http://www.ft.com (Accessed 6 April 2008).

Beattie, A. (2010) 'Brazil to Suspend Action in US Cotton Dispute', *Financial Times*, 18 June, http://www.ft.com (Accessed 22 June 2010).

Beattie, A. and Blas, J. (2008) 'Precious Grains', *Financial Times*, 14 April. http://www.ft.com (Accessed 14 April 2008).

Beddies, S., Correia, M., Kolavalli, S. and Townsend, R. (2006) 'Reform of Coffee and Cotton Crop Boards in Tanzania', in A. Coudouel, A. Dani, and S. Paternostro (eds) *Poverty and Social Impact Analysis of Reforms* (Washington, DC: World Bank).

Bendaña, A. (2006) *NGOs and Social Movements: A North/South Divide?* Civil Society and Social Movements Programme Paper No. 22 (Geneva: UNRISD).

Bendaña, A. (2008) 'From Development Assistance to Development Solidarity: The Role of Venezuela and ALBA', Paper presented at the conference on 'The Changing Face of Global Development Finance – Impacts and implications for Aid, Development, the South and the Bretton Woods Institutions', Ottawa.

Benicchio, R. (2005) 'The WTO ruling on Brazil-US Cotton: Implications for African Countries and Agriculture Negotiations', in E. Hazard (ed.) *International Trade Negotiations and Poverty Reduction: The White Paper on Cotton* (Dakar: Enda Diapol), 33–42.

Berman, B. (1992) 'Up from Structuralism', in B. Berman and J. Lonsdale *Unhappy Valley: Conflict in Kenya & Africa: Book One: State & Class* (London: James Currey).

Bernstein, S. and Cashore, B. (2007) 'Can Non-State Global Governance be Legitimate? An Analytical Framework', *Regulation and Governance*, 1, 1–25.

Berry, S. (1993) *No Condition is Permanent: The Social Dynamics of Agrarian Change in Sub-Saharan Africa* (Madison: University of Wisconsin Press).

Bhagwati, J. (1958) 'Immiserizing Growth: A Geometrical Note', *Review of Economic Studies*, 25: 3, 201–205.

Bhagwati, J. and Ruggie, J.G. (eds) (1984) *Power, Passions, and Purpose: Prospects for North-South Negotiations* (Cambridge, MA: MIT Press).

Bienefeld, M. (1982) 'Tanzania: Model or Anti-Model?', in M. Bienefeld and M. Godfrey (eds) *The Struggle for Development: National Strategies in an International Context* (New York: John Wiley).

Bienefeld, M. (1986) 'Analyzing the Politics of African State Policy: Some Thoughts on Robert Bates' Work', *IDS Bulletin*, 17: 1, 5–11.

Biermann, W. and Wagao, J. (1986) 'The Quest for Adjustment: Tanzania and the IMF, 1980–1986', *African Studies Review*, 29: 4, 89–103.

bioRe Tanzania (2006) *Internal Control System (ICS) Manual* (Mwamishali: bioRe Tanzania).

Birdsall, N. (2007) 'Stormy Days on an Open Field: Asymmetries in the Global Economy', in G. Mavrotas and A. Shorrocks (eds) *Advancing Development: Core Themes in Global Economics* (Basingstoke: Palgrave Macmillan and UNU-WIDER), 227–251.

Blas, J. (2007) 'Political Sensitivity over Food Price Rises', *Financial Times Special Report: World Economy 2007*, 8 October, 13.

Blas, J. (2008) 'UN Warns of Food "Neo-Colonialism"', *Financial Times*, 19 August.

Blouin, C. and Weston, A. (2005) 'Agriculture at the WTO: A Development Test for the Doha Round', *NSI Review*, Fall-Winter 2005–2006.

Blowfield, M. (2007) 'Reasons to be Cheerful? What We Know about CSR's Impact', *Third World Quarterly*, 28: 4, 683–695.

Bond, P. (2005) 'Neoliberalism in Sub-Saharan Africa: Structural Adjustment to NEPAD', in A. Saad-Filho and D. Johnston (eds) *Neoliberalism: A Critical Reader* (London: Pluto Press), 230–236.

Borowiak, C. (2004) 'Farmers' Rights: Intellectual Property Regimes and the Struggle over Seeds', *Politics and Society*, 32: 4, 511–543.

BRIDGES Weekly Trade News Digest (2007a) 'WTO High-Level Meeting on Cotton Exceeds Low Expectations', 11: 10, 21 March 2007.

BRIDGES Weekly Trade News Digest (2007b) 'WTO Members Welcome AG Chair's Paper, but Criticise "Balance"', 11: 16, 9 May 2007.

BRIDGES Weekly Trade News Digest (2007c) 'After Long Delay, US Notifies 2002–2005 AG Subsidies to WTO', 11: 34, 10 October 2007.

BRIDGES Weekly Trade News Digest (2007d) 'Draft Texts Facing Delay', 11: 40, 21 November 2007.

BRIDGES Weekly Trade News Digest (2008a) 'US Appeals Compliance Ruling', 12: 6, 20 February 2008.

BRIDGES Weekly Trade News Digest (2008b) 'US Senate Passes Farm Bill', 12: 1, 16 January 2008.

BRIDGES Weekly Trade News Digest (2008c) 'US Farm Bill Talks Heat Up', 12: 5, 13 February 2008.

BRIDGES Weekly Trade News Digest (2008d) 'LDCs Outline Priorities', 12: 8, 5 March 2008.

BRIDGES Weekly Trade News Digest (2008e) 'US, Brazil Clash on Cotton Sanctions', 12: 6, December.

BRIDGES Weekly Trade News Digest (2010) 'Brazil, US Strike "Framework" Deal in Cotton Dispute', 14: 23, June.

Bryceson, D.F. (1982). 'Peasant Commodity Production in Post-Colonial Tanzania', *African Affairs*, 81: 325, 547–567.

Bryceson, D.F (1997) 'De-agrarianisation: Blessing or Blight?', in D.F. Bryceson and V. Jamal (eds) *Farewell to Farms: De-agrarianisation and Employment in Africa* (Aldershot: Ashgate), 237–256.

Campbell, B. (1975) 'Neo-Colonialism, Economic Dependence and Political Change: Cotton Textile Production in the Ivory Coast', *Review of African Political Economy*, 2: 2, 36–53.

Cashore, B. (2002) 'Legitimacy and the Privatization of Environmental Governance: How Non- State Market Driven (NSMD) Governance Systems Gain Rule Making Authority', *Governance*, 15: 4, 503–529.

Cerny, P.G. (2000) 'Structuring the Political Arena: Public Goods, States and Governance in a Globalizing World', in R. Palan (ed.) *Global Political Economy: Contemporary Theories* (London: Routledge), 21–35.

Chachage, C.S.L. (1993) 'Forms of Accumulation, Agriculture and Structural Adjustment in Tanzania', in P. Gibbon (ed.) *Social Change and Economic Reform in Africa* (Uppsala: The Scandinavian Institute of African Studies).

Chang, H-J. (2008) *Bad Samaritans: The Myth of Free Trade and the Secret History of Capitalism* (New York: Bloomsbury Press).

The Citizen [Dar es Salaam] (2008) 'Tanzania: Passbook Loaning System Scrapped', 21 June 2008. http:www.allafrica.com (Accessed on 26 August 2008).

Clapp, J. and Fuchs, D. (eds) (2009) *Corporate Power in Global Agrifood Governance* (Cambridge, Massachusetts: The MIT Press).

CMIA (Cotton Made in Africa) (2006) Various pages. http://www.cotton-made-in-africa.com (Accessed 12 September 2007).

Coleman, W.D. (2005) 'Uruguay Round', in W.D. Coleman and N. Johnson (eds) *Globalization and Autonomy Online Compendium*. http://www.globalautonomy. ca/global1/glossary_entry.jsp?id=EV.0020 (Accessed 31 March 2008).

Coleman, W.D., Grant, W. and Josling, T. (2004) *Agriculture in the New Global Economy* (London: Edward Elgar).

Coleman, W.D. and Reed, A.J. (2006) 'Legalization, Transnationalism and Organic Agriculture', in W.D. Coleman and N. Johnson (eds) *Globalization and Autonomy Online Compendium*. http://www.globalautonomy.ca (Accessed 4 January 2008).

Coleman, W.D. and Wayland S. (2006) 'The Origins of Global Civil Society and Nonterritorial Governance: Some Empirical Reflections', *Global Governance*, 12: 241–261.

Collier, P. (2007) *The Bottom Billion: Why the Poorest Countries are Failing and What Can be Done About It* (Oxford: Oxford University Press).

CRB (Commodity Research Bureau) (2004) *The CRB Commodity Yearbook* (New York: John Wiley).

Conroy, M. (2007) *Branded! How the 'Certification Revolution' is Transforming Global Corporations* (Gabriola Island, Canada: New Society Publishers).

Cooksey, B. (2003) 'Marketing Reform? The Rise and Fall of Agricultural Liberalization in Tanzania', *Development Policy Review*, 21: 1, 67–91.

Cooksey, B. (2004) *Elixir or Poison Chalice? The Relevance of Aid to East Africa*, Paper presented at the OECD DAC Development Partnership Forum on Improving Donor Effectiveness in Combating Corruption, Paris.

Cornia, G.A., Jolly, R. and Stewart, F. (1987) *Adjustment with a Human Face* (Oxford: Oxford University Press).

Cox, R.W. (1996a) 'Ideologies and the New International Economic Order: Reflections on Some Recent Literature', in R.W. Cox with T.J. Sinclair (eds) *Approaches to World Order* (Cambridge: Cambridge University Press), 376–419.

Cox, R.W. (1996b) 'Social Forces, States and World Orders: Beyond International Relations Theory', in R.W. Cox with T.J. Sinclair (eds) *Approaches to World Order* (Cambridge: Cambridge University Press), 85–123.

Crooks, E. and Green, M. (2007) 'Soaring Fuel Bills Put Pressure on Africa's Fragile Economies', *Financial Times*, 29/30 December, 3.

Culpeper, R. (2006) 'Creating Fiscal Space through Improved Domestic Resource Mobilization in Low-Income Countries', Paper presented at the FONDAD/UN-DESA Roundtable on Policy Space in Developing Countries, New York.

Cutler, A.C., Haufler, V. and Porter, T. (eds) (2001) *Private Authority and International Affairs* (New York: SUNY Press).

de Man, R. (2006) *Promoting Sustainable Cotton Production in West Africa: Potential Supply Chain Strategies. Report to UNEP and FAO.* http://www.rdeman.nl (Accessed 15 June 2009).

Dercon, S. (1993) 'Peasant Supply Response and Macroeconomic Policies: Cotton in Tanzania', *Journal of African Economies*, 2: 157–194.

Derr, J. (2007) 'RSE et le Développement: Un Accord Gagnant-Gagnant? L'Exemple du PPP "Cotton made in Africa" Mémoire' (Germany: Leipzig University).

Dietler, C. and Guntern, J. (2005) 'Organic Cotton for Poverty Alleviation? Project Evaluation Mali, Burkina Faso and Kyrgyzstan', Paper presented to the symposium 'Combining Consumer's Concern and Poverty Reduction', Lucerne.

Diouf, E.H. and Hazard, E. (2005) 'Sectoral Initiative on Cotton: A Balancing Act of Alliances in Africa and at the WTO', in E. Hazard (ed.) *International Trade Negotiations and Poverty Reduction: The White Paper on Cotton* (Dakar: Enda Diapol), 57–69.

Diouf, J. and Severino, J.M. (2008) 'Africa Must Grow to Rely on Its Own Farms', *The Guardian Weekly*, 2 May 2008, 18.

Doane, D. (2005) 'The Myth of CSR: The Problem with Assuming that Companies Can Do Well While also Doing Good is that Markets Don't Really Work that Way', *Stanford Social Innovation Review*, Fall.

Dollar, D. and Kray, A. (2001) *Trade, Growth and Poverty*, World Bank Policy Research Working Paper No. 2615 (Washington, DC: World Bank).

Dorward, A., Kydd, J. and Poulton, C. (eds) (1998) *Smallholder Cash Crop Production under Market Liberalization* (New York: CAB International).

Drache, D. (2006) *Trade, Development and the Doha Round: A Sure Bet or a Train Wreck?* Robarts Centre Research Papers (Toronto: York University).

Drache, D. (2008) *Defiant Publics: The Unprecedented Reach of the Global Citizen* (London: Polity Press).

EAC (East African Community) (2007) *East African Organic Standard* [Final Draft] (Arusha: EAC Regional Standards Technical Working Group).

The East African (2005) 'Tanzania to Conduct Field-Trials of GM Cotton', 28 March 2005. http://allafrica.com (Accessed 10 November 2007).

The East African (2008a) 'Outgrowers to Boost Cotton Production', 14 July 2008. http://allafrica.com (Accessed 28 August 2008).

The East African (2008b) 'Tanzania: Donors to Loosen Purse Strings for Dar', 6 October 2008. http://allafrica.com (Accessed 8 October 2008).

Easterly, W. (2006) *The White Man's Burden: Why The West's Efforts to Aid the Rest Have Done So Much Ill and So Little Good* (New York: The Penguin Press).

ECA (Economic Commission for Africa) (2005) *The Millennium Development Goals in Africa: Progress and Challenges* (Addis Ababa: ECA).

ECA-AU (2008) *Economic Report on Africa 2008: Africa and the Monterrey Consensus: Tracking Performance and Progress* (Addis Ababa: EAC and AU).

The Economist (2007) 'Trade Talks Mangling Trade', 7 July 2007, 86.

Elhiraika A. and Ndikumana, L. (2007) 'Reserves Accumulation in African Countries: Sources, Motivations and Effects', Mimeo (Amherst, MA: Department of Economics, University of Massachusetts).

England, A. (2005) 'Agriculture: Crucial Sector Needs Help to Blossom', *Financial Times Special Report: Tanzania*, 2 August.

EPOPA (Export Promotion of Organic Products from Africa) (2006) *Organic Exporter Quide: Hands-On Help for Organic Exports from Africa* (AB Bennekom, The Netherlands: EPOPA).

Evans, A. and Ngalwea, E. (2003) 'Tanzania', *Development Policy Review*, 21: 2, 271–287.

Evans, P. (1995). *Embedded Autonomy: States and Industrial Transformation* (Princeton: Princeton University Press).

Eyhorn, F., Ratter, S.O. and Ramakrishnan, M. (2005) *Organic Cotton Crop Guide: A Manual for Practitioners in the Tropics* (Frick, Switzerland: Research Institute of Organic Agriculture).

Fairtrade Foundation (2006) 'Questions and Answers about Fairtrade Certified Cotton'. http: www.fairtrade.org.uk/ (Accessed 7 July 2007).

FLO (Fairtrade Labelling Organization International) (2006) *Standards du Commerce Équitable pour le Coton Graine pour les Organisations de Petits Producteurs* (London: FLO).

FAO (Food and Agriculture Organization) (2004a) *The State of Agricultural Commodity Markets 2004* (Rome: Commodities and Trade Division).

FAO (2004b) *Cotton No. 1: Impact of Support Policies on Developing Countries: A Guide to Contemporary Analysis*, FAO Trade Policy Technical Notes (Rome: FAO).

FAO (2004c) *Cotton No. 1: Impact of Support Policies on Developing Countries: Why do the Numbers Vary?* FAO Trade Policy Briefs No. 1 (Rome: FAO).

Farnie, D.A. and Jeremy, D.J. (eds) (2004) *The Fibre that Changed the World: The Cotton Industry in International Perspective, 1600s–1990s* (Oxford: Oxford University Press).

Ferrigno, S., Ratter, S.G., Ton, P., Vodouhe, D.S., Williamson, S. and Wilson, J. (2005) *Organic Cotton: A New Development Path for African Smallholders*, Gatekeeper Series 120 (London: Natural Resources Group, International Institute for Environment and Development).

Feyerabend, P. (1975) *Against Method* (London: Humanities Press).

Financial Times (2007) 'Spinning Cotton', 25 October, 14.

Financial Times (2010) 'US Must Stand Up for Free Trade – Cotton', 22 June, 8.

Finnemore, M. and Sikkink, K. (1998) 'International Norm Dynamics and Political Change', *International Organization*, 52: 4, 887–917.

Fjeldstad, O., Kolstad, I. and Nygaard, K. (2006) *Bribes, Taxes and Regulations: Business Constraints for Micro Enterprises in Tanzania*, CMI Working Paper 2006: 2 (Bergen, Norway: Chr. Michelsen Institute).

The Foundation for Civil Society (2006) *Annual Report 2005* (Dar es Salaam: The Foundation for Civil Society).

Forum-Coton (2004) 'Cotton Sub-Sector Reforms in Tanzania', *Forum Union Européenne – Afrique sur le coton*, Paris: 5–6 juillet 2004. http://www.forum-coton.org/docs/presentations/2.4-en.pdf (Accessed 8 March 2008).

Friedman, M. (1970) 'The Social Responsibility of Business is to Increase Its Profits', *The New York Times Magazine*, 13 September 1970.

Friedman, T. (2000) *The Lexus and the Olive Tree: Understanding Globalization*, Reprint edn (New York: Anchor Books).

Galbraith, J.K. (1967) *The New Industrial State* (Boston: Houghton Mifflin Company).

Galbraith, J.K. (1973) *Economics and the Public Purpose* (Boston: Houghton Mifflin Company).

Galbraith, J.K. (1979) *The Nature of Mass Poverty* (Harvard: Harvard University Press).

Gallagher, K.P. (2008) 'Understanding Developing Country Resistance to the Doha Round', *Review of International Political Economy*, 15: 1, 62–85.

Gereffi, G. and Korzeniewicz, M. (eds) (1994) *Commodity Chains and Global Capitalism* (Westport, CT: Greenwood Press).

Gibbon, P. (1998) *King Cotton under Market Sovereignty: The Private Marketing Chain for Cotton in Western Tanzania, 1997/98*, Centre for Development Research Working Paper no. 98.17 (Copenhagen: Centre for Development Research).

Gibbon, P. (1999) 'Free Competition without Sustainable Development? Tanzanian Cotton Sector Liberalization, 1994/95 to 1997/98', *Journal of Development Studies*, 36: 1, 128–150.

Gibbon, P. (2001) 'Cooperative Cotton Marketing, Liberalization and "Civil Society", in Tanzania', *Journal of Agrarian Change*, 1: 3, 389–439.

Gibbon, P. (2007) 'Africa, Tropical Commodity Policy and the WTO Doha Round', *Development Policy Review*, 25: 1, 43–70.

Gibbon, P., Havnevik, K.J. and Hermele, K. (1993) *A Blighted Harvest: The World Bank and African Agriculture in the 1980s* (London: James Currey and Africa World Press).

Gibbon, P. and Ponte, S. (2005) *Trading Down: Africa, Value Chains and the Global Economy* (Philadelphia: Temple University Press).

Giles, C. (2007) 'Poll Reveals Backlash in Wealthy Countries against Globalization: Majority Want Higher Taxation of the Rich', *Financial Times*, 23 July, 1.

Gillson, I., Poulton, C., Balcombe, K. and Page, S. (2004) 'Understanding the Impact of Cotton Subsidies on Developing Countries', Working paper. May.

Girvan, N. (2007) 'Power Imbalances and Development Knowledge', Theme Paper prepared for the project 'Southern Perspectives on Reform of the International Development Architecture' (Ottawa: The North-South Institute. http://www.nsi-ins.ca (Accessed 4 March 2008).

Gladwin, C.H. (1991) 'Introduction', in C.H. Gladwin (ed.) *Structural Adjustment and African Women Farmers* (Gainesville: University of Florida Press).

Global Policy Forum (2006) 'Total UN System Estimated Expenditures'. http://www.globalpolicy.org/finance/tables/system/tabsyst.htm (Accessed 3 July 2006).

Goodwin, B.K. and Mishra, A.K. (2006) 'Are "Decoupled" Farm Program Payments Really Decoupled? An Empirical Evaluation', *American Journal of Agricultural Economics*, 88: 1, 73–89.

Goreux, L. (2004) 'Cotton after Cancún', Draft Discussion Paper for the OECD Sahel and West African Club, March 2004.

Goreux, L. (2005) 'Cut Subsidies, Beware Diversionary Tactics and Don't Miss the Hong Kong Window of Opportunity', in E. Hazard (ed.) *International Trade Negotiations and Poverty Reduction: The White Paper on Cotton* (Dakar: Enda Diapol).

Gouse, M., Shankar, B. and Thirtle, C. (2008) 'The Decline of Bt Cotton in KwaZulu-Natal: Technology and Institutions', in W.G. Moseley and L.C. Gray (eds) *Hanging by a Thread: Cotton, Globalization, and Poverty in Africa* (Ohio: Ohio University Press), 103–121.

GRAIN (2003) 'GM Cotton Set to Invade West Africa'. http://www.grain.org (Accessed 4 November 2005).

Green, R.H. (1974) 'Tanzania', in H. Chenery et al. (eds) *Redistribution with Growth* (Oxford: World Bank and Institute for Development Studies), 268–273.

Harrison, G. (2001) 'Post-Conditionality Politics and Administrative Reform: Reflections on the Cases of Uganda and Tanzania', *Development and Change*, 32: 4, 657–679.

Hårsmar, M. (2004) *Heavy Clouds but No Rain: Agricultural Growth Theories and Peasant Strategies on the Mossi Plateau, Burkina Faso*. Doctoral dissertation, Department of Rural Development Studies, Swedish University of Agricultural Sciences, Uppsala.

Havnevik, K.J. (1993) *Tanzania: The Limits to Development from Above* (Uppsala: The Nordic Africa Institute).

Hazard, E. (2001) *Commerce des Produits Agricoles et Lutte Contre la Pauvreté, le Cas du Coton Ouest Africain: Le Coton Ouest Africain: Une Production Ingegree au Commerce Mondial et Compétitive* (Dakar: ENDA Diapol).

Hazard, E. (2002) *Farm Trade and Poverty Reduction: West African Cotton* (Dakar: ENDA Diapol).

Hazard, E. (ed.) (2005) *Le Livre Blanc sur le Coton: Négociations Commerciales Internationales et Réduction de la Pauvreté*, 2nd edn (Dakar: ENDA Tiers Monde).

Helleiner, E. (1994) *States and the Reemergence of Global Finance* (Ithaca: Cornell University Press).

Helleiner, G.K. (1981) 'The Refsnes Seminar: Economic Theory and North-South Negotiations', *World Development*, 9: 6, 539–555.

Helleiner, G.K. (1987) 'Stabilization, Adjustment and the Poor', *World Development*, 15: 12, 1499–1513.

Helleiner, G.K. (1999) 'The Legacies of Julius Nyerere: An Economist's Reflections', Paper prepared for the Queen's University Studies in National and International Development Series panel on Julius Nyerere, Kingston.

Helleiner, G.K. (2000) 'Markets, Politics and Globalization: Can the Global Economy be Civilized?' Paper delivered at the United Nations Conference on Trade and Development 10th Raul Prebisch Lecture, Palais des Nations, Geneva.

Helleiner, G.K. (2006) 'Trade and Trade-related Policies for Development and Poverty Reduction: The Capacity Building Challenge', Paper presented at the International Lawyers and Economists Against Poverty Conference, Arusha.

Helleiner, G.K., Killick, T., Lipumba, N., Ndulu, B.J. and Svendsen, S.K. (1995) *Report of the Group of Independent Advisors on Development Cooperation Issues between Tanzania and Its Donors'* (Dar es Salaam: Royal Danish Ministry of Foreign Affairs).

Hillocks, R.J. (2005) 'Is There a Role for Bt Cotton in IPM for Smallholders in Africa?', *International Journal of Pest Management*, 51: 2, 131–141.

Hobson, J.M. (2004) *The Eastern Origins of Western Civilization* (Cambridge: Cambridge University Press).

Hoekman, B. and Kostecki, M. (2001) *The Political Economy of the World Trading System: The WTO and Beyond*, 2nd edn (Oxford: Oxford University Press).

Hopkins, M. (2005) Measurement of Corporate Social Responsibility', *International Journal of Management and Decision Making*, 6: 3–4, 213–231.

Hormeku, T. (1998) 'NGOs Critical of UNCTAD's Investment and Pro-Business Approach', *Third World Economics*, 187/188: 1.

Hosea, K.M., Msaki, O.N. and Swai, F. (2004) *Genetically Modified Organisms (GMOs) in Tanzania* (Dar es Salaam: Envirocare Tanzania).

Hull, I.V. (2003) 'Military Culture and the Production of "Final Solutions" in the Colonies: The Example of Wilhelminian Germany', in R. Gellately and B. Kiernan (eds) *The Specter of Genocide: Mass Murder in Historical Perspective* (Cambridge: Cambridge University Press), 141–162.

Hulme, D. (2008) 'Reflections on NGOs and Development: The Elephant, the Dinosaur, Several Tigers but No Owl', in A.J. Bebbington et al. (eds) *Can NGOs Make a Difference? The Challenge of Development Alternatives* (London: Zed Books), 337–345.

Huntington, S. (1968) *Political Order in Changing Societies* (New Haven: Yale University Press).

Hyden, G. (2006) *African Politics in Comparative Perspective* (Cambridge: Cambridge University Press).

Hymer, S.H. and Resnick, S.A. (1970) 'International Trade and Uneven Development', in J. Bhagwati, R.W. Jones, R.A. Mundell and J. Vanek (eds) *Trade, Balance of Payments and Growth* (Amsterdam: North Holland).

Ibrahim Index of African Governance (2008) http://www.moibrahimfoundation.org (Accessed 16 August 2008).

ICAC (International Cotton Advisory Committee) (1978) Minutes: 291st Meeting of the Standing Committee (SC-M-291) (Washington, DC: ICAC).

ICAC (1982) Minutes: 314th Meeting of the Standing Committee (SC-M-314), (Washington, DC: ICAC).

ICAC (2004) *Cotton: World Statistics*, November 2004. http://www.icac.org (27 April 2009).

ICAC (2007) 'Rapport de la Réunion Inaugurale du Panel d'Experts sur la Performance Sociale, Environnementale et Économique (SEEP) de la Production Cotonnière'. http://www.icac.org (Accessed 4 January 2008).

ICAC and Estur, G. (2007) 'Africa's Cotton in the World', Paper presented to a seminar on cotton in Africa held on 6–8 September at Arusha, Tanzania. http://www.icac.org (Accessed 30 October 2007).

Iliffe, J. (1972) 'Tanzania under German and British Rule', in L. Cliffe and J. Saul (eds) *Socialism in Tanzania: Politics* (Nairobi: East African Publishing House), 8–17.

Imboden, N. (2004) 'Société Civile et OMC: Rôle et Place des Entités Extra-Gouvernementales dans la Definition des Politiques Commerciales et la Formulation des Positions de Négociations Africaines', Paper presented to the ICTSD Round Table, Geneva.

Inter Press Service (2008) 'Food Crisis Linked to Doha Deal', 8 May 2008. http://www.ipsnews.net (Accessed 9 May 2008).

Irwin, D.A. (1995) 'The GATT in Historical Perspective', *The American Economic Review*, 85: 2, 323–328.

Irwin, D.A. (2003) 'The Optimal Tax on U.S. Cotton Exports', *Journal of International Economics*, 60: 2, 275–291.

Isaacman, A.F. (1976) *The Tradition of Resistance in Mozambique: Anti-colonial Activity in the Zambesi Valley, 1850–1921* (Berkeley: University of California Press).

Isaacman, A.F. (1993) 'Peasants and Rural Social Protest in Africa', in F. Cooper, A.F. Isaacman, F.E. Malon, W. Roseberry and S.J. Stern (eds) *Confronting Historical Paradigms* (Madison: University of Wisconsin Press).

Isaacman, A.F. (1996) *Cotton is the Mother of Poverty: Peasants, Work and Rural Struggle in Colonial Mozambique, 1938–1961* (Portsmouth: Heinemann).

Isaacman, A.F. and Roberts, R. (eds) (1995) *Cotton, Colonialism and Social History in Sub-Saharan Africa* (Portsmouth: Heinemann).

Isaacman, A.F., Stephen, M., Adam, Y., Joao Homen, M., Macamo, E. and Pililao, A. (1980) 'Cotton as the Mother of Poverty: Peasant Resistance to Forced Cotton Production in Mozambique, 1938–1961', *International Journal of African Historical Studies*, 13: 581–615.

Jayne, T.S., Mather, D. and Mghenyi, E. (2006) *Smallholder Farming under Increasingly Difficult Circumstances: Policy and Public Investment Priorities for Africa*, MSU International Development Working Paper No. 86 (East Lansing: Michigan State University).

Jewsiewicki, B. (1980) 'African Peasants in the Totalitarian Colonial Society of the Belgian Congo', in M. Klein (ed.) *Peasants in Africa: Historical and Contemporary Perspectives* (Beverly Hills: Sage), 45–76.

Jolly, R. (2007) 'Inequality in Historical Perspective', in G. Mavrotas and A. Shorrocks (eds) *Advancing Development: Core Themes in Global Economics* (Basingstoke: Palgrave Macmillan and UNU-WIDER), 63–73.

Jomo, K.S. (2005) 'Introduction', in K.S. Jomo (ed.) *The Long Twentieth Century: The Great Divergence: Hegemony, Uneven Development, and Global Inequality* (New Delhi: Oxford University Press), 1–24.

Jomo, K.S. and Fine, B. (eds) (2006) *The New Development Economics* (London: Zed Books).

Jones, W.O. (1987) 'Food-Crop Marketing Boards in Tropical Africa', *Journal of Modern African Studies*, 25: 3, 375–402.

Kabelwa, G. and Kweka, J. (2006) *The Linkage Between Trade, Development and Poverty Reduction: A Case Study of Cotton and Textile Sector in Tanzania* [Draft] (Dar es Salaam: Economic and Social Research Foundation).

Kajoki, J. (2008) 'CCM in Crash Appeal for Nyanza Cooperative Body', *The Citizen*, 23 May 2008. http://www.thecitizen.co.tz/ (Accessed 23 August 2008).

Kamat, S. (2004) 'The Privatization of Public Interest: Theorizing NGO Discourse in a Neoliberal Era', *Review of International Political Economy*, 11: (1) 155–176.

Kanbur, R. (2001) 'Economic Policy, Distribution, and Poverty: The Nature of Disagreements', *World Development*, 29: 1083–1094.

Kanbur, R. (2003) *Conceptual Challenges in Poverty and Inequality: One Development Economist's Perspective* (Ithaca: Cornell University).

Kanaan, O. (2000) 'Tanzania's Experience with Trade Liberalization', *Finance and Development*, 37: 2. http://www.imf.org (Accessed 18 March 2008).

Kaplinsky, R. (2005) *Globalization, Poverty and Inequality* (Cambridge: Polity Press).

Karplus, V.J. and Deng, X.W. (2006) 'Agricultural Biotechnology in China: Past, Present and Future', in X. Zhihong, L. Jiayang, X. Yongbiao, W. Yang (eds) *Biotechnology and Sustainable Agriculture 2006 and Beyond* (Netherlands: Springer).

Kaufmann, J. (ed.) (1989) *Effective Negotiations: Case Studies in Conference Diplomacy* (The Hague: Martinus Nijoff Publishers and United Nations Institute for Training and Research).

Keck, M.E. and Sikkink, K. (1998) 'Transnational Advocacy Networks in International Politics: Introduction', in M.E. Keck and K. Sikkink (eds) *Activists Beyond Borders: Advocacy Networks in International Politics* (Ithaca: Cornell University Press).

Keohane, R.O. (1984) *After Hegemony: Cooperation and Discord in the World Political Economy* (Princeton: Princeton University Press).

Keynes, J.M. (1933) 'National Self-Sufficiency', *The Yale Review*, 22: 4, 755–769.

Khagram, S., Riker, J.V. and Sikkink, K. (eds) (2002) 'From Santiago to Seattle: Transnational Advocacy Groups Restructuring World Politics', in *Restructuring World Politics: Transnational Movements, Networks and Norms* (Minneapolis: University of Minnesota Press), 3–23.

Khan, K-ur-R. (1982) *The Law and Organization of International Commodity Agreements* (The Hague: Martinus Nijhoff Publishers).

Khan, M.H. (2006) 'Corruption and Governance', in K.S. Jomo and B. Fine (eds) *The New Development Economics* (London: Zed Books), 200–220.

Kherallah, M. (2002) *Reforming Agricultural Markets in Africa* (Baltimore: Johns Hopkins University Press).

Khor, M. (2007) 'The Recent WTO Agricultural Negotiations: Domestic Support', Paper presented at the UNDP regional trade workshop 'Doha and Beyond', Penang, Malaysia, December.

Kiely, R. (2007) 'Poverty Reduction Through Liberalization? Neoliberalism and the Myth of Global Convergence', *Review of International Studies*, 33, 415–434.

Kiondo, A.S.Z. (1993) 'Structural Adjustment and Nongovernmental Organizations in Tanzania: A Case Study', in P. Gibbon (ed.) *Social Change and Economic Reform in Africa* (Uppsala: The Scandinavian Institute of African Studies).

Krasner, S.D. (ed.) (1983) *International Regimes* (Ithaca: Cornell University Press).

Kripke, G. (2005) 'King Cotton: Abdicate or Subjugate?', in E. Hazard (ed.) *International Trade Negotiations and Poverty Reduction: The White Paper on Cotton* (Dakar: Enda Diapol).

Krugman, P.R. (1984) 'Import Protection as Export Promotion', in H. Kierzkowski (ed.) *Monopolistic Competition and International Trade* (Oxford: Clarendon Press).

Lamsdorff, J. (2007) *Corruption Perceptions Index 2006*. http://www.transparency. org/ (Accessed 4 January 2008).

Larsen, M.N. (2003) *Quality Standard-Setting in the Global Cotton Chain and Cotton Sector Reforms in Sub-Saharan Africa*, Working Paper 03.1 Institute for International Studies (Copenhagen: Institute for International Studies).

Larsen, M.N. (2004) *Governing Post-Liberalised Markets: National Market Coordination and the Global Cotton Chain*, Doctoral dissertation, Institute of Geography, Faculty of Science, University of Copenhagen, Denmark, August 2004.

Larsen, M.N. (2006) 'Market Coordination and Social Differentiation: A Comparison of Cotton-Producing Households in Tanzania and Zimbabwe', *Journal of Agrarian Change*, 6: 1, 102–131.

Lavelle, K. (2001) 'Ideas within a Context of Power: The African Group in an Evolving UNCTAD', *The Journal of Modern African Studies*, 39: 1, 25–50.

Lee, D. (2009) 'Bringing an Elephant into the Room: Small African State Diplomacy in the WTO', in A.F. Cooper & T.M. Shaw (eds) *The Diplomacies of Small States: Between Vulnerability and Resistance* (Houndmills, Basingstoke: Palgrave Macmillan), 195–206.

Lele, U., van de Walle, N. and Gbetibouo, M. (1989) *Cotton in Africa: An Analysis of Differences in Performance*, MADIA Discussion Paper No. 7 (Washington: World Bank).

Lewis, W.A. (1965) *Development Planning: The Essentials of Economic Policy* (London: George Allen & Unwin).

Leys, C. (1975) *Underdevelopment in Kenya: The Political Economy of Neo-Colonialism, 1964–1971* (London: James Currey).

Leys, C. (1996) *The Rise and Fall of Development Theory* (Oxford: James Currey).

Li, Y. and Rowe, F. (2007) *Aid Inflows and the Real Effective Exchange Rate in Tanzania Policy*, Research Working Paper WPS4456 (Washington, DC: World Bank).

Lipton, M. (2005) *The Family Farm in a Globalizing World: The Role of Crop Science in Alleviating Poverty*, IFPRI 2020 Discussion Paper No. 40 (Washington, DC: IFPRI).

Lipton, M. and Ravallion, M. (1993) *Poverty and Policy*, Policy Research Working Paper Series No. 1130 (Washington, DC: World Bank).

Löfgren, M. and Thörn, H. (2007) 'Introduction: Global Civil Society – More or Less Democracy?', *Development Dialogue*, 49: 5–15.

Maizels, A. (2003) 'Economic Dependence on Commodities', in J. Toye (ed.) *Trade and Development: Directions for the 21st Century* (Cheltenham: Edward Elgar), 169–184.

Mamdani, M. (1996) *Citizen and Subject: Contemporary Africa and the Legacy of Late Colonialism* (Princeton: Princeton University Press).

May, J. (2001) 'An Elusive Consensus: Definitions, Measurement and Analysis of Poverty', Paper presented at the Economic Commission for Africa expert group meeting on pro-poor growth strategies for Africa, June.

Meena, R. (1991) 'The Impact of Structural Adjustment Programs on Rural Women in Tanzania', in C.H. Gladwin (ed.) *Structural Adjustment and African Women Farmers* (Gainesville: University of Florida Press), 169–190.

Milanovic, B. (2002) *The Two Faces of Globalization: Against Globalization As We Know It*, World Bank Research Working Paper (Washington, DC: World Bank).

Milanovic, B. (2003) *Is Inequality in Africa Really Different?* World Bank Research Working Paper WPS3169 (Washington, DC: World Bank).

Mitlin, D., Hickey, S. and Bebbington, A. (2007) 'Reclaiming Development? NGOs and the Challenge of Alternatives', *World Development*, 35: 10, 1699–1720.

Mkandawire, T. and Olukoshi, A. (eds) (1995) *Between Liberalisation and Oppression: The Politics of Structural Adjustment in Africa* (Dakar: CODESRIA).

Moseley, W.G. (2005) 'Global Cotton and Local Environmental Management: The Political Ecology of Rich and Poor Small-Hold Farmers in Southern Mali', *The Geographical Journal*, 171: 1, 36–55.

Moseley, W.G. and Gray, L.C. (eds) (2008) *Hanging By a Thread: Cotton, Globalization, and Poverty in Africa* (Ohio: Ohio University Press).

Msuya, E. (2007) *The Impact of FDI on Agricultural Productivity and Poverty Reduction in Tanzania*, Munich Personal RePEc Archive Paper (Kyoto: Kyoto University).

Mulley, S. (2006) *Tanzania: A Genuine Case of Recipient Leadership in the Aid System* (London: UK Aid Network).

Murphy, C. (1984) *Emergence of the NIEO Ideology* (Boulder, CO: Westview Press).

Myrdal, G. (1970) *The Challenge of World Poverty: A World Anti-Poverty Program in Outline* (New York: Random House).

Mytelka, L.K. (1989) 'The Unfulfilled Promise of African Industrialization', *African Studies Review*, 32: 3, 77–137.

Nafziger, E.W. (1986) 'Review of C.K. Eicher and D.C. Baker "Research on Agricultural Development in Sub-Saharan Africa: A Critical Survey"', *Economic Development and Cultural Change*, 34: 4, 876–881.

Narayan, D. (1997) *Voices of the Poor: Poverty and Social Capital in Tanzania* (Washington, DC: World Bank).

Narayan, D., Patel, R., Schafft, K., et al. (2000) *Voices of the Poor: Can Anyone Hear Us?* (Washington, DC: World Bank).

Narlikar, A. and Tussie, D. (2004) 'The G20 at the Cancún Ministerial: Developing Countries and Their Evolving Coalitions in the WTO', *The World Economy*, 27: 7, 947–966.

Ndikumana, L. (2000) 'Financial Determinants of Domestic Investment in Sub-Saharan Africa: Evidence from Panel Data', *World Development*, 28: 2, 381–400.

Nevitte, N. (1996) *The Decline of Deference: Canadian Value Change in Cross National Perspective* (Peterborough: Broadview Press).

Newell, P. and Frynas, J.G. (2007) 'Beyond CSR? Business, Poverty and Social Justice: An Introduction', *Third World Quarterly*, 28: 4, 669–681.

Nicholls, A. and Opal, C. (2005) *Fair Trade: Market-Driven Ethical Consumption* (London: Sage Publications).

North, D. (1990) *Institutions, Institutional Change and Economic Performance* (Cambridge: Cambridge University Press).

Nurkse, R. (1953) *Problems of Capital Formation in Underdeveloped Countries* (Oxford: Blackwell).

Nyerere, J. (1967) *Freedom and Socialism* (Oxford: Oxford University Press).

Nyerere, J. (1973) *Freedom and Development* (Oxford: Oxford University Press).

OAU (Organization for African Unity) (1980) *Lagos Plan of Action: 1980–2000*. http://www.uneca.org/ (Accessed 1 August 2008).

Ocaya-Lakidi, D. (1977) 'Manhood, Warriorhood and Sex in East Africa', in A.A. Mazrui (ed.) *The Warrior Tradition in African Politics* (Boston: BRILL).

O'Connor, D. (2007) 'Policy Lessons for 21st Century Industrializers', in *Industrial Development for the 21st Century: Sustainable Development Perspectives* (New York: UN DESA), 415-422.

ODI (Overseas Development Institute) (2004) *Developed Country Cotton Subsidies and Developing Countries: Unravelling the Impacts on Africa*, ODI Briefing Paper (London: ODI). http://www.odi.org.uk (Accessed 10 November 2005).

OECD (Organization for Economic Cooperation and Development) (2005) *Paris Declaration on Aid Effectiveness: Twelve Indicators of Progress* (Paris: Development Co-operation Directorate – Development Assistance Committee).

OECD (2006) *Cotton in West Africa: The Economic and Social Stakes* (Paris: OECD Publishing).

OECD (2008) *Accra Agenda for Action* (Accra: 3rd High Level Forum on Aid Effectiveness).

Ohmae, K. (1990) *The Borderless World* (New York: Harper Business).

Onyeiwu, S. (2000) 'Deceived by African Cotton: The British Cotton Growing Association and the Demise of the Lancashire Textile Industry', *African Economic History*, 28: 89–121.

Organic Exchange (2007) *Organic Cotton Market Report 2007* (O'Donnell, TX: Organic Exchange).

Organic Exchange (2008) *Organic Farm and Fibre Report 2008* (O'Donnell, TX: Organic Exchange).

Ostry, S. (2000) 'The Multilateral Trading System', in A.M. Rugman and T.L. Brewer (eds) *Oxford Handbook of International Business*, 1: 9, 232–259.

Owen, E.R.J. (1969) *Cotton and the Egyptian Economy, 1820–1914: A Study in Trade and Development* (Oxford: Clarendon Press).

Oyejide, T.A. (2004) 'Costs and Benefits of Special and Differential Treatment for Developing Countries in GATT/WTO', in T.A. Oyejide and W. Lyakurwa (eds) *Africa and the World Trading System: Selected Issues of the Doha agenda*, vol. 1 (Trenton, NJ: Africa World Press and the African Economic Research Consortium), 175–211.

Ozden, C. and Reinhardt, E. (2003) *The Perversity of Preferences: The Generalized System of Preferences and Developing Country Trade Policies, 1976–2000*, World Bank Policy Research Working Paper No. 2955 (Washington, DC: World Bank).

PAN UK (Pesticide Action Network UK) (2006) 'Rachel Carson Memorial Lecture 2006: Farmers and Fashion: From Harvest to High Street'. http://www.pan-uk.org (1 September 2007).

Peuples Solidaires (2002) 'West Africa – Let Us Save Cotton'. http://www.peuples-solidaires.org/article123.html (Accessed 3 July 2006).

Pesche, D. and Nubukpo, K. (2005) 'The African Cotton Set in Cancún: A Look Back at the Beginning of Negotiations', in E. Hazard (ed.) *International Trade Negotiations and Poverty Reduction: The White Paper on Cotton* (Dakar: Enda Diapol), 45–55.

Pinder, J. (1998) *The Building of the European Union*, 3rd edn (Oxford: Oxford University Press).

Polanyi, K. (1957) 'The Economy as Instituted Process', in K. Polanyi, C.M. Arensberg and H.W. Pearson (eds) *Trade and Market in the Early Empires* (Glencoe: The Free Press).

Politi, J. and Wheatley, J. (2010a) 'Brazil Moves Closer to Showdown Over US Cotton Subsidies', *Financial Times*, 8 March, 4.

Politi, J. and Wheatley, J. (2010b) 'Tax Move by Brazil Risks US Trade War', *Financial Times*, 9 March, 1.

Porter, M.E. and Kramer, M.R. (2006) 'Strategy and Society: The Link between Competitive Advantage and Corporate Social Responsibility', *Harvard Business Review*, 1 December 2006, 78–92.

Poulton, C. (1998) 'Cotton Production and Marketing in Northern Ghana: The Dynamics of Competition in a System of Interlocking Transactions', in A. Dorward, J. Kydd and C. Poulton (eds) *Smallholder Cash Crop Production under Market Liberalization: A New Institutional Economics Perspective* (New York: CAB International), 56–112.

Poulton, C. (2006) *Multi-Country Review of the Impact of Cotton Sector Reform in Sub-Saharan Africa: Tanzania Country-Study Zero Draft* (London: Centre for Environmental Policy and World Bank).

Poulton, C., Gibbon, P., Hanyani-Mlambo, E., Kydd, J., Maro, W., Nylandsted Larsen, M., Tschirley, D. and Zulu, B. (2004a) *Competition and Coordination in Cotton Market Systems of Southern and Eastern Africa*, Research Report: DFID Project R8080 (London: DFID).

Poulton, C., Gibbon, P., Hanyani-Mlambo, E., Kydd, J., Maro, W., Nylandsted Larsen, M., Tschirley, D. and Zulu, B. (2004b) 'Competition and Coordination in Liberalized African Cotton Market Systems', *World Development*, 32: 3, 519–536.

Poulton, C., Maro, W. and Nylandsted Larsen, M. (2005) 'Competition and Coordination in the Tanzanian Cotton Sector 2001–2004'. www.aec.msu.edu/fs2/zambia/Tanzania_cotton_workshop.pdf (Accessed 9 June 2009).

Pratt, C. (2000) 'Julius Nyerere: The Ethical Foundation of His Legacy', *The Round Table*, 355: 365–374.

Prebisch, R. (1950) *The Economic Development of Latin America and Its Principal Problems* (New York: United Nations).

Putnam, R. (1988) 'Diplomacy and Domestic Politics: The Logic of Two-Level Games', *International Organization*, 42: 427–461.

The RATES Centre (2003) *Cotton-Textile-Apparel: Value Chain Report Tanzania* (Nairobi: Kenya).

Ratter, S.G. (2004) 'Organic and Fair Trade Cotton in Africa', Paper presented to the Organic Cotton Conference, Hamburg, February 2004.

Ratter, S.G. (2005) *Organic Cotton Production in Tanzania Improves Food Security* (Pähl, Germany: bioSim).

Raynolds, L., Murray, D. and Wilkinson, J. (2007) *Fair Trade: The Challenges of Transforming Globalization* (London: Routledge).

Ravallion, M. and Chen, S. (2008) *The Developing World is Poorer than We Thought, but No Less Successful in the Fight Against Poverty*, World Bank Research Working Paper WPS4703 (Washington, DC: World Bank).

Reilly-King, F. and Sneyd, A. (eds) (2008) *The Changing Face of Global Development Finance – Impacts and Implications for Aid, Development, the South and the Bretton Woods Institutions* (Ottawa: Halifax Initiative).

Reich, R. (2007) *Supercapitalism: The Transformation of Business, Democracy and Everyday Life* (New York: Vintage).

Reuters (2010) 'Africa Left in Cold by US-Brazil Cotton Deal: Study', 15 April. http://af.reuters.com (Accessed 15 April 2010).

Richardson, B. (2009) *Sugar: Refined Power in a Global Regime* (Houndmills, Basingstoke: Palgrave Macmillan).

Richey, L.A. and Ponte, S. (2008) 'Better (Red)™ than Dead? Celebrities, Consumption and International Aid', *Third World Quarterly*, 29: 4, 711–729.

Rockefeller Foundation (2006) *Africa's Turn: A New Green Revolution for the 21ˢᵗ Century* (New York: The Rockefeller Foundation).

Rodney, W. (1972) *How Europe Underdeveloped Africa* (Washington, DC: Howard University Press).

Rodrik, D. (2007) *One Economics, Many Recipes: Globalization, Institutions, and Economic Growth* (Princeton: Princeton University Press).

Ruggie, J.G. (1982) 'International Regimes, Transactions and Change: Embedded Liberalism in the Post-War Economic Order', *International Organization*, 36: 2, 379–415.

Rutasitara, L. (2002) *Economic Policy and Rural Poverty in Tanzania* (Dar es Salaam: Mkuki na Nyota Publishers).

Sachs, J.D. (2005) *The End of Poverty: Economic Possibilities for Our Time* (New York: Penguin Press).

Sagasti, F., Bezanson, K. and Prada, F. (2005) *The Future of Development Financing: Challenges and Strategic Choices* (Basingstoke: Palgrave Macmillan).

Sandbrook, R. (1982) *The Politics of Basic Needs: Urban Aspects of Assaulting Poverty in Africa* (London: Heinemann).

Saul, J.S. (1973) 'Marketing Cooperatives in Tanzania', in L. Cliffe and J.S. Saul (eds) *Socialism In Tanzania*, vol. 2 (Nairobi: East African Publishing House), 141–153.

Saul, J.S. (1974) 'The State in Post-Colonial Societies: Tanzania', in R. Miliband and S. Saville (eds) *Socialist Register 1974*, 11: 349–372. http://www.socialistregister.com (Accessed 21 July 2008).

Scholte, J.A. (2002) 'Civil Society and Governance in the Global Polity', in M. Ougaard and R. Higgott (eds) *Towards a Global Polity* (London: Routledge).

Scholte, J.A. (2005) *Globalisation: A Critical Introduction*, 2ⁿᵈ edn (Basingstoke: Palgrave Macmillan).

Scholte, J.A. (2007) 'Global Civil Society – Opportunity or Obstacle for Democracy?', *Development Dialogue*, 49: 15–29.

Scholte, J.A. with Schnabel, A. (eds) (2002) *Civil Society and Global Finance* (London: Routledge).

Schumpeter, J.A. (1950) *Capitalism, Socialism and Democracy* (New York: Harper & Brothers).

Sen, A. (1999) *Development as Freedom* (New York: Anchor Books).

Shadlen, K.C. (2005) 'Exchanging Development for Market Access? Deep Integration and Industrial Policy under Multilateral and Regional-Bilateral Trade Agreements', *Review of International Political Economy*, 12: 5, 750–775.

Shao, J.R. (2002) *Agriculture and Market Liberalization in Tanzania: Problems of Cotton Production and Marketing in Bunda District*, Tanzania Agriculture Situation Analysis Report (Dar es Salaam: Tanzania Development Research Group).

Shaw, L. (1995) 'Producer Consumer Cooperation in Cotton', Paper presented to the UNCTAD Standing Committee on Commodities, Geneva.

Shepherd, A. and Farolfi, S. (1999) *Export Crop Liberalization in Africa: A Review*, FAO Agricultural Services Bulletin No. 135 (Rome: FAO).

Sikkink, K. (2002) 'Restructuring World Politics: The Limits and Asymmetries of Soft Power', in Khagram et al. (eds) *Restructuring World Politics* (Minneapolis: University of Minnesota Press), 301–317.

Singer, H.W. (1950) 'The Distribution of Gains from Trade between Investing and Borrowing Countries', *American Economic Review*, 40: 1950, 473–485.

Singer, H.W. and Ansari, J.A. (1992) *Rich and Poor Countries*, 4th edn (London: Routledge).

Sneyd, A. (2003) *Globalizing Embedded Liberalism: Some Lessons for the WTO's Development Round*, Robarts Centre Research Paper (Toronto: York University). http://www.yorku.ca/robarts/projects/wto/pdf/wto_globalembedliberal.pdf (Accessed 30 March 2008).

Sneyd, A. (2005) 'Commodity Trade', in W.D. Coleman and N. Johnson (eds) *Globalization and Autonomy Online Compendium*. http://www.globalautonomy.ca (Accessed 4 March 2008).

Sneyd, A. (2006a) 'Terms of Trade', in W.D. Coleman and N. Johnson (eds) *Globalization and Autonomy Online Compendium*. http://www.globalautonomy.ca (Accessed 4 March 2008).

Sneyd, A. (2006b) 'North-South', in W.D. Coleman and N. Johnson (eds) *Globalization and Autonomy Online Compendium*. http://www.globalautonomy.ca (Accessed 30 March 2008).

Sneyd, A. (2006c) 'The Wade-Wolf Debate on Global Inequality', in W.D. Coleman and N. Johnson (eds) *Globalization and Autonomy Online Compendium*. http://www.globalautonomy.ca (Accessed 4 March 2008).

Sneyd, A. (2007) *Empowering Southern Knowledge to Enable Southern Ownership* (Report on Wilton Park Conference WP887, Southern Perspectives on Reform of the International Development Architecture Project) (Ottawa: The North-South Institute). http://www.nsi-ins.ca/english/pdf/WiltonParkConferenceReport.pdf (Accessed 4 March 2008).

Soros, G. (2008) *The New Paradigm for Financial Markets: The Credit Crisis of 2008 and What It Means* (New York: Public Affairs).

Steffek, J. (2006) *Embedded Liberalism and Its Critics: Justifying Global Governance in the American Century* (Basingstoke: Palgrave Macmillan).

Stern, D.L. (2008) 'In Tajikistan, Debt-Ridden Farmers Say They are Pawns', *The New York Times*, 15 October 2008.

Stern, S. (2007) 'Goodbye to Corporate Social Responsibility?', *Financial Times*, 31 December, 9.

Stewart, F. (2003) 'Income Distribution and Development', in J. Toye (ed.) *Trade and Development: Directions for the 21st Century* (Oxford: Oxford University Press), 185–217.

Stewart, F. (2007) 'Do We Need a New "Great Transformation"? Is One Likely?', in G. Mavrotas and A. Shorrocks (eds) *Advancing Development: Core Themes in Global Economics* (Basingstoke: Palgrave Macmillan and UNU-WIDER), 614–639.

Stewart, F., Laderchi, C. and Saith, R. (2007) 'Introduction: Four Approaches to Defining and Measuring Poverty', in F. Stewart, R. Saith, and B. Harriss-White (eds) *Defining Poverty in the Developing World* (Basingstoke: Palgrave Macmillan), 1–35.

Stiglitz, J.E. (1989) 'Markets, Market Failures, and Development', *The American Economic Review*, 79: 2, 197–203.

Stiglitz, J.E. (2006a) *Making Globalization Work* (New York: W.W. Norton).

Stiglitz, J.E. (2006b) 'The Tyranny of King Cotton', *The Guardian*, 24 October 2006. http://www.guardian.co.uk/ (Accessed 13 August 2008).

Stiglitz, J. and Charlton, A. (2005) *Fair Trade for All: How Trade Can Promote Development* (Oxford: Oxford University Press).

Strange, S. (1994) 'Who Governs? Networks of Power in World Society', *Hitotsubashi Journal of Law and Politics*, Special edn, 5–17.

Sumner, D. (2007) *US Farm Programs and African Cotton*, IPC Issue Brief 22 (Washington, DC: International Food and Agricultural Trade Policy Council).

TACOGA (Tanzania Cotton Growers Association) (2006) Letter to M/S Holden Ginners Ltd. Re: Seed Cotton Price and Passbook to Farmers, 14 August 2006, Mwanza.

Talbott, I.D. (1990) *Agricultural Innovation in Colonial Africa: Kenya and the Great Depression* (New York: Edwin Mellen Press).

Tallontire, A. (2007) 'CSR and Regulation: Towards a Framework for Understanding Private Standards Initiatives in the Agri-Food Chain', *Third World Quarterly*, 28: 4, 775–791.

Taylor, I. (2003) 'Global Monitor: The United Nations Conference on Trade and Development', *New Political Economy*, 8: 3, 409–418.

Taylor, I. (2006) *China and Africa: Engagement and Compromise* (London: Routledge).

TCB (Tanzania Cotton Board) (2007) *Corporate Strategic Plan: 2007/08–2009/10* (Dar es Salaam: TCB).

Thal Larsen, P. (2007) 'Super-Rich Widen the Wealth Gap by Taking Risks', *Financial Times*, 28 June, 13.

Tchirley, D.L., Poulton, C., Gergely, N., Labaste, P., Baffes, J., Boughton, D. and Estur, G. (2010) 'Institutional Diversity and Performance in African Cotton Sectors', *Development Policy Review*, 28: 3, 295–323.

Thrasher, R.D. and Gallagher, K.P. (2008) *21ˢᵗ Century Trade Agreements: Implications for Long-Run Development Policy*, The Pardee Papers, No. 2 (Boston University: Pardee Centre).

TNDRDP (Tanzania-Netherlands District Rural Development Programme) (2004) *Baseline Survey on Poverty, Welfare and Services in Rural Shinyanga District* (Bukoba: Economic Development Initiatives).

Tomlinson, B. (2008) 'World Aid Trends: Donors Distorting the Reality of Aid in 2008', in *The Reality of Aid 2008: Aid Effectiveness: Democratic Ownership and Human Rights* (Quezon City, Philippines: IBON Books).

Toye, J. (1994) 'Structural Adjustment: Context, Assumptions, Origin and Diversity', in R. van der Hoeven and F. van der Kraaij (eds) *Structural Adjustment and Beyond in Sub-Saharan Africa* (London and The Hague: James Currey and Ministry of Foreign Affairs), 18–35.

Toye, J. and Toye, R. (2003) 'The Origins and Interpretations of the Prebisch-Singer Thesis', *History of Political Economy*, 35: 3, 437–467.

Toye, J. and Toye, R. (2004) 'World Monetary Problems and the Challenge of Commodities', in *The UN and Global Political Economy: Trade, Finance and Development* (Indiana: University of Indiana Press), 230–275.

Trebilcock, M.J. and Howse, R. (1999) *The Regulation of International Trade*, 2ⁿᵈ edn (London: Routledge).

UNCTAD (United Nations Conference on Trade and Development) (2004a) *The Least Developed Countries Report 2004: Linking International Trade with Poverty Reduction*, UNCTAD/LDC/2004 (New York and Geneva: United Nations).

UNCTAD (2004b) *The São Paulo Consensus* UNCTAD/TD/410 (New York and Geneva: United Nations).

UNCTAD (2007) *Economic Development in Africa: Reclaiming Policy Space – Domestic Resource Mobilization and Developmental States*, UNCTAD/ALDC/AFRICA/2007/1 (New York and Geneva: UNCTAD).

UNCTAD (2008) *World Investment Directory: Volume X Africa.* http://www.unctad. org (2 September 2008).

UNCTAD-ICC (2005) *Investment Guide: Tanzania.* http://www.unctad.org/Templates/ WebFlyer.asp?intItemID=3449&lang=1 (Accessed 9 June 2009).

UNDP (United Nations Development Programme) (1980) *Supplementary Assistance for a Global Project: Integrated Cotton Research and Development Programme*, DP/PROJECTS/R.13/Add.6 (New York: UNDP).

UNDP (2009) *Human Development Report 2009: Tanzania: The Human Development Index.* http://hdrstats.undp.org/en/countries/country_fact_sheets/cty_fs_ TZA.html. (Accessed 9 November 2009).

UNEP-UNCTAD (2008) *Organic Agriculture and Food Security in Africa*, UNEP-UNCTAD Capacity-building task force on trade, environment and development (New York and Geneva: UNEP and UNCTAD).

UNESCO (United Nations Educational, Scientific and Cultural Organization) (2008) 'Regional Initiatives: APROCA: University of Cotton', *International Association of Universities E-Bulletin*, 5: 2, March. http://www.unesco.org (Accessed 10 June 2008).

UNGC-UNDP (2004) *Growing Sustainable Business for Poverty Reduction in Tanzania: Summary of Related Initiatives* (New York: UNDP).

UNMP (United Nations Millennium Project) (2005) *Investing in Development: A Practical Plan to Achieve the Millennium Development Goals* (London and Sterling VA: Earthscan).

United Nations (1978) 'United States of America and United Republic of Tanzania: Project Agreement Relating to the Arusha Region Drought Assistance Project', *UN Treaty Series No. 17239.* Signed at Dar es Salaam on 12 and 13 August 1975.

United Nations (2004) 'Paradigm Shifts Advocated by the Panel', in *We The Peoples: Civil Society, the United Nations and Global Governance: Report of the Panel of Eminent Persons on United Nations-Civil Society Relations*, A/58/817 (Geneva: United Nations).

United Nations Industrial Development Organization (UNIDO) (2005) *UNIDO-Scope*, 29 August–14 September. http://www.unido.org (Accessed on 22 April 2008).

United Nations Negotiating Conference on a Common Fund under the Integrated Programme for Commodities (1980) *Agreement Establishing the Common Fund for Commodities* (Geneva: United Nations).

URT (United Republic of Tanzania) (1997) *Mwanza Region Socio-Economic Profile* (Dar es Salaam & Mwanza: The Planning Commission and Mwanza Regional Commissioner's Office).

URT (2004a) *Tanzania FDI Report 2004* (Dar es Salaam: Government Printer).

URT (2004b) *Vulnerability and Resilience to Poverty in Tanzania: Causes, Consequences and Policy Implications: 2002/03 Tanzania Participatory Poverty Assessment* (Dar es Salaam: Research and Analysis Working Group).

URT (2005a) *The Environmental Management Act* (Act Supplement No. 3) (Dar es Salaam: Government Printer).

URT (2005b) *The National Biosafety Framework for Tanzania* (Dar es Salaam: Vice President's Office Division of the Environment).

URT (2005c) *National Strategy for Growth and Reduction of Poverty (NSGRP) [MKUKUTA]* (Dar es Salaam: Vice President's Office).

URT (2006) *Government Circular on the Reform of the Financing, Structures Roles and Responsibilities of Crop Boards* (Dar es Salaam: Ministry of Agriculture, Food Security and Cooperatives).

URT (2007) *Views of the People 2007: Tanzanians Give their Opinions on Growth and Reduction of Income Poverty, their Quality of Life and Social Well-Being, and Governance and Accountability* (Dar es Salaam: RAWG, MKUKUTA Monitoring System).

USDA (United States Department of Agriculture) (2002) *Agricultural Outlook: August* (Washington: Economic Research Service).

Utting, P. (2007) 'CSR and Equality', *Third World Quarterly*, 28: 4, 697–712.

van der Hoeven, R. and van der Kraaij, F. (1994) 'Discussions and Conclusions', in R. van der Hoeven and F. van der Kraaij (eds) *Structural Adjustment and Beyond in Sub-Saharan Africa* (London and The Hague: James Currey and Ministry of Foreign Affairs), 174–185.

van de Walle, N. (2001) *African Economies and the Politics of Permanent Crisis, 1979–1999* (Cambridge: Cambridge University Press).

Vogel, D. (2005) *The Market for Virtue: The Potential and Limits of CSR* (Washington, DC: Brookings Institution Press).

Wade, R.H. (2004) *Governing the Market: Economic Theory and the Role of Government in East Asian Industrialization*, 2nd paperback edn, with a new introduction by the author (Princeton: Princeton University Press).

Wade, R.H. (2005) 'Why Free Trade has Costs for Developing Countries', *Financial Times*, 10 August.

Wade, R. H. (2008) 'Globalization, Growth, Poverty, Inequality, Resentment, and Imperialism', in J. Ravenhill (ed.), 2nd edn, *Global Political Economy* (Oxford: Oxford University Press), 373–409.

Wangwe, S.M. and Semboja, H.H. (2002) 'Impact of Structural Adjustment on Industrialization and Technology in Africa', in T. Mkandawire and C.C. Soludo (eds) *African Voices on Structural Adjustment* (Ottawa: IDRC/CODESRIA/Africa-World Press).

Watkins, K. and Fowler, P. (2002) 'Primary Commodities: Trading into Decline', in *Rigged Rules and Double Standards: Trade, Globalization and the Fight against Poverty* (Oxford: Oxfam). http://www.maketradefair.com (Accessed 28 February 2006).

Watkins, K. with Sul, J-u. (2002) *Cultivating Poverty: The Impact of US Cotton Subsidies on Africa*, Oxfam Briefing Paper 30 (Oxford: Oxfam International).

Watkins, M. (2003) 'Politics in the Time and Space of Globalization', in W. Clement and L.F. Vosko (eds) *Changing Canada: Political Economy as Transformation* (Montreal and Kingston: McGill-Queen's University Press).

Weiss, L. (2005) 'Global Governance, National Strategies: How Industrialized States Make Room to Move under the WTO', *Review of International Political Economy*, 12: 5, 723–749.

Weston, A. (2007) 'Remembering the World's Poorest Cotton Growers', *Review* (Ottawa: The North-South Institute), 1–2 and 10.

Wiarda, H.J. (1982) 'Cancún and After: The United States and the Developing World', *PS*, 15: 1, 40–48.

Williams, F. (2008) 'UN Drive for West to Focus Attention on Economic and Social Rights', *Financial Times*, 9 January 2008, 8.

Williamson, S., Ferrigno, S. and Vodouhe, S.D. (2005) 'Needs-Based Decision-Making for Cotton Problems in Africa: A Response to Hillocks', *International Journal of Pest Management*, 51: 4, 219–224.

Winham, G.R. (1984) *International Trade and the Tokyo Round Negotiation* (Princeton: Princeton University Press).

WIPO (World Intellectual Property Organization) (2007) 'Championing the Fight Against Fake Drugs', *WIPO Magazine*, January 2007. http://www.wipo.int (Accessed 1 September 2008).

Witt, H., Patel, R. and Schnurr, M. (2006) 'Can the Poor Help GM Crops? Technology, Representation and Dotton in the Makhathini Flats, South Africa', *Review of African Political Economy*, 109: 497–513.

Wolcott, S. (2005) 'Review of D.A. Farnie and D.J. Jeremy (eds) *"The Fibre That Changed the World: The Cotton Industry in International Perspective*, 1600–1990s"', *The Journal of Economic History*, 65: 2, 604–606.

Wolf, M. (2005a) 'How to Help Africa Escape Poverty Trap', *Financial Times*, 11 January.

Wolf, M. (2005b) 'The Elimination of Poverty', *Financial Times*, 15 February.

Wolf, M. (2007) 'Why Plutocracy Endangers Emerging Market Economies', *Financial Times*, 7 November, 11.

Wolfe, R. (1998) *Farm Wars: The Political Economy of Agriculture and the International Trade Regime* (New York: St. Martin's Press).

Woods, N. (2006) 'Mission Unaccomplished in Africa', in *The Globalizers: The IMF, the World Bank and Their Borrowers* (Ithaca: Cornell University Press), 141–178.

World Bank (1981) *Accelerated Development in Sub-Saharan Africa* [*The Berg Report*] (Washington, DC: World Bank).

World Bank (2007) *World Development Report 2008: Agriculture for Development*, (Washington, DC: World Bank).

World Bank (2008) *Africa Development Indicators 2007* (Washington, DC: World Bank).

Worsham III, J.B. (2003) 'The Mystery of the Micronaire – How It is Costing You Money', *Cotton Grower*, Reprinted with permission from Meister Publishing, April 2003. http://www.cottoninc.com (Accessed 1 September 2008).

WTO (World Trade Organization) (2000) *Annual Report 2000* (Geneva: WTO).

WTO (2001) *Annual Report 2001* (Geneva: WTO).

WTO (2003a) *Poverty Reduction: Sectoral Initiative in Favour of Cotton. Joint Proposal by Benin, Burkina Faso, Chad and Mali*, 16 May 2003. TN/AG/GEN/4.

WTO (2003b) 'Address by President Blaise Compaoré of Burkina Faso', 10 June 2003. http://www.wto.org (Accessed 6 April 2006).

WTO (2003c) 'Wording of Paragraph 27 of the Revised Draft Cancún Ministerial Text', 7 October 2003, WT/GC/W/516.WTO (2004a) *WTO African Regional Workshop on Cotton*, 23–24 March, WT/L/564. http://www.wto.org (Accessed 6 April 2008).

WTO (2004b) *Text of the July Package – The General Council's Post-Cancun Decision*, WT/L/579. http://www.wto.org (Accessed 6 April 2008).

WTO (2004c) *Upland Cotton* (Report of the Panel United States Upland Cotton), 8 September 2004, WT/DS267/R8 (Geneva: WTO).

WTO (2005a) *Upland Cotton* (Report of the Appellate Body United States Upland Cotton), 3 March 2005, WT/DS267/AB/R3 (Geneva: WTO).

WTO (2005b) *Hong Kong Ministerial Declaration*, WT/MIN(05)/DEC. http://www.wto.org (Accessed 6 April 2008).

WTO (2006) *Director-General's Consultative Framework Mechanism on Cotton Development Assistance: Evolving Working Table*, 15 December 2006, WT/L/670 (Geneva: WTO).

WTO (2007a) 'Director-General's Summary Remarks', *Director-General's Consultative Framework Mechanism on Cotton: High Level Session*, 15–16 March 2007, TN/AG/SCC/W/7 (Geneva: WTO).

WTO (2007b) *Director-General's Consultative Framework Mechanism on Cotton Development Assistance: Evolving Table*, 19 November 2007, WT/L/702 (Geneva: WTO).

Yaro, J.A. (2006) 'Is Deagrarianisation Real? A Study of Livelihood Activities in Rural Northern Ghana', *The Journal of Modern African Studies*, 44: 125–156.

Zachary, G.P. (2007) 'Out of Africa: Cotton and Cash', *New York Times*, 14 January.

Zadek, S. (2005) 'Responsible Competitiveness: Reshaping Global Markets Through Responsible Business Practices', *Corporate Governance*, 6: 4, 334–348.

Zunckel, H.E. (2005) 'The African Awakening in United States – Upland Cotton', *Journal of World Trade*, 39: 6, 1071–1109.

Index